Learning from the Land

Learning from the Land

Teaching Ecology through Stories and Activities

Second Edition

Brian "Fox" Ellis

Illustrated by Vin Luong

LIBRARIES UNLIMITED

AN IMPRINT OF ABC-CLIO, LLC
Santa Barbara, California • Denver, Colorado • Oxford, England

Library of Congress Cataloging-in-Publication Data

Ellis, Brian, 1962–
    Learning from the land : teaching ecology through stories and activities / Brian "Fox" Ellis ; illustrated by Vin Luong.
        p. cm.
    Includes bibliographical references and index.
    ISBN 978-1-59884-918-9 (hardcopy : alk. paper) — ISBN 978-1-59884-919-6 (ebook)
1. Environmental sciences—Study and teaching—Activity programs.   2. Storytelling.
I. Title.
    GE77.E45   2012
    372.35′7044—dc23   2011026443

ISBN: 978-1-59884-918-9
EISBN: 978-1-59884-919-6

16   15   14   13   12   1   2   3   4   5

This book is also available on the World Wide Web as an eBook.
Visit www.abc-clio.com for details.

Libraries Unlimited
An Imprint of ABC-CLIO, LLC

ABC-CLIO, LLC
130 Cremona Drive, P.O. Box 1911
Santa Barbara, California 93116-1911

This book is printed on acid-free paper ∞

Manufactured in the United States of America

I would like to dedicate this book to the people of the Earth,
those who take the time to listen to the blooming of flowers,
the songs of the water, and the cacophony of frogs in the spring.
I would also like to dedicate this book to you, the reader.
May you learn these stories, write a few of your own,
and tell them to the future stewards of the Earth.

Contents

Preface to the New Edition

As I was finishing my rewrite and revision of this book, I took a break to meet some good friends for a cup of tea and conversation. These are not just friends but mentors, heroes of mine. Joan and Bob Ericksen have spent their lives creating opportunities for young people to be immersed in the wild world through the arts. They founded The Sun Foundation more than 30 years ago. They hire me every summer for a week-long camp called "Arts and Science in the Woods." I am on a committee with them that organizes "The Clean Water Celebration" in which 3,000 students from throughout the state of Illinois come together to learn about our local watershed through hands-on science, storytelling, art, and music. (We coauthored a "How-To" book available for free from my website: http://www.foxtalesint.com/Books/CleanWaterCelebrationHow-ToBook) They also helped me to find funding for my documentary, "Voices for the River."

Joan has this wonderful habit of buying a case of any book she really loves and giving them away to all her friends. She gave me a book by David Suzuki called *The Legacy* and she said that I had to read it. I thanked her and told her that I was wrapping up a writing project and I would get to it eventually. I decided to save it as a reward for finishing this book.

Our conversation turned to the idea of legacy: How will this excellent life work of theirs continue as they move toward retirement? How has their effort impacted the next generation of children, "those whose faces have not yet pushed up from the earth," as Seneca Elder Corn Planter once said?

They in turn challenged me to think about how stories shape future generations. I told them the story of a recent chance encounter: I was hiking at Starved Rock State Park when a young man in his early 20s stopped me and asked if I was that storyteller who visited his school for a week-long residency when he was in the fourth grade. He then went on to tell me how stories I had taught him more than a dozen years ago had inspired him to become a summer camp counselor and staff naturalist. He used some of the stories in *Learning from the Land* at camp last summer. I have to admit that this chance encounter not only made my day, my month, my year—it made my whole life feel somehow more worthwhile.

The Stories Themselves Have a Life of Their Own Passed from One Generation to Another

I read David Suzuki's book that night, cover to cover. It is wonderful, challenging, and inspiring. I should warn you that he earns the nickname Margaret Atwood chides him with, "Dr. Doom and Gloom." The first few chapters are filled with dark and depressing statistics about what is wrong with the environment, how human behavior has poisoned our waterways, erased species, and changed the climate. It is worth wading through because of his ability to look past the problem and offer real insight into the causes. Do not look away, because it is vital that we fully understand the root of the problem before we can begin to solve it.

He creates hope and inspiration with solutions that are within our grasp. Yes, small things do add up, if everyone changed to fluorescent bulbs, or better yet LED, we could reduce greenhouse gases in measureable ways. If we recycle, reduce our over-consumptive habits, reuse and re-purpose technology, eat locally, and vote with our dollars, we can make a difference.

But big changes are needed and these changes are rooted in empathy, changing the way we think about other creatures, seeing ourselves as part of the earth and its cycles, understanding those cycles and how we can harmonize within them.

Modern science offers great insight, as do the stories of Native Elders from many earth-based cultures the world over. He says that the stories we tell are the key to this new understanding.

Suzuki ends his eloquent and compassionate book by saying that we need a new story: "I am uplifted by the amazing story emerging from modern science." And later writes, "we lost our way, forgot the narrative that reminded us of who we are, why we are here and where we belong." He says the solutions are found in these narratives and goes on to quote philosopher Thomas Berry:

> Tell me the story of the river and the valley and the streams and woodlands and wetlands, of shellfish and finfish. A story of where we are and how we got here and the characters and the roles we play. Tell me a story, a story that will be my story as well as the story of everyone and everything about me, the story that brings us together in a valley community, a story that brings together the human community with every living being in the valley, a story that brings us together under the arc of the great blue sky in the day and the starry heavens at night. *(Thomas Berry.* The Dream of the Earth. *San Francisco: Sierra Club Books, 1988.)*

This book is my humble attempt to tell those stories, and more than that, this book is my effort to inspire you to tell your story. Many of these stories were originally written more than 20 years ago, maybe ahead of their time, but they have found a voice in the environmental educators who use these stories in their programs, and in the classroom teachers who use these lessons to inspire their students to send a voice.

Allow the quiet whisper in a field of prairie flowers to be heard, and become a voice for these songs.

May these stories and lessons inspire the next generation to harmonize with the song of the cosmos, the spiral dance of the Milky Way, the whirl of the milkweed seed within the pod waiting for the winds of autumn to push us toward the next unfolding of spring.

A Note on Some of the Changes in this New Version

I was honored and more than a little excited when my publisher asked me to revise and update *Learning from the*

Land. Clearly the book has done well and I hope it has served its readers, helping them find and tell the stories of their local landscape.

One of the most difficult, but most useful changes was in reformatting all of the lesson plans. You will find a more step-by-step outline, National Standards, timelines, and grade level cues.

Fifteen years ago I had researched websites for many of the lesson plans, but my editor at the time said teachers do not use these resources and they are too ephemeral with URLs changing all of the time, so I cut them from the original edition, and then added them back into this update.

I also reorganized the bibliography. Instead of a long list of books at the end of this one, I have added books and websites into each chapter, including only the titles that I found truly useful and ones that will help educators with the lesson plans. There has been a plethora of great news books about the environment for younger readers!

I have also added three new stories, plus several poems and songs. Two of these lesson plans first appeared as article in the magazine *Green Teacher*, which I highly recommend to any educator, formal or informal, who wishes to expand his or her knowledge of the environment and find classroom-tested ideas for teaching ecological issues: http://www.greenteacher.com/

Since this book was first published I have done extensive work on bringing natural history to life by portraying historical naturalists, including John James Audubon, Charles Darwin, and Gregor Mendel. I have included one of Audubon's stories as well as some lesson plans to help you step into character.

I have also founded a theater company and written a fair amount of musical theater, so there is also a lesson on taking old songs and rewriting them as environmental sing-a-longs.

So this revised and expanded edition of *Learning from the Land* reflects both my growth as a performer and educator as well as some of the scientific breakthrough and technological advances we have made as a culture. I hope you find it useful, and please let me know what you think: www.foxtalesint.com

Telling Earth Tales,

Brian "Fox" Ellis

Preface to the First Edition

Working in my garden, weeding the asparagus patch, I came across a small quartz arrowhead and my mind reeled with the possibilities. I saw a Cherokee boy stalking a deer in the shadows of towering oaks. I heard the rhythm of two stones clacking together as a craftsman molded the tools of survival. Beyond that, I heard the crash of two continents adrift, the grinding and shattering of rocks, molten silica seeping into the cracks and slowly cooling, crystallizing, to form the quartz stone I held in my hand.

Everywhere I travel on the earth the rocks sing. Tales stalk me from behind trees and hillsides. The legends of this land are clamoring to be heard.

Stories of the Land

In the essay "Good, Wild, Sacred," Gary Snyder tells of the journey tales of the Australian Aborigines. These stories serve as maps that bring this arid desert region to life with mythical creatures and human adventures. Each watering hole holds a story. Each prominent rock reminds the travelers of their location and reminds them of their relationship to the land. These stories are told from one generation to another. If people know the stories of the land, they are never lost. The people feel a deeper connection with their personal and cultural history through the landscape.

The Klamath tell similar stories that contain the history of their people and the best fishing holes along the river. The Navajo landscape is said to stalk the people; cliff faces hold stories that remind the people of appropriate behavior within the community and appropriate relationships with the thunder-beings and creepy crawlies.

Indigenous people around the world have enchanted successive generations with the stories of the land. Some of these stories have been lost to changes in the landscape and changes in the dreamscape of our rapidly changing culture. But still the hills and elders provide deep reservoirs of story. If you seek these stories, you will find them.

Learning the Stories of the Land

How does one learn the stories of the land?

As a storyteller, I have conditioned myself to listen. I have passed long hours squatting on a stone in the middle of Rock Creek allowing the music of the dancing water to filter into my thoughts, drinking the water and allowing it to flow into my veins. I began to learn the songs of water.

You can do the same. Listen to the elders; listen to the songs of the wild world around you.

Befriend an elderly person in your neighborhood and the history of your home will unfold in conversation. Spend the day working in their garden and they will fill your ear with tales tall and true. Ask your neighbors or check the county clerk's office to learn about the recent history of your land; read up on local geology and natural history as well as the people who previously walked this Earth.

A few weeks ago, I went to a sawmill with an elderly friend to get sawdust for his goat barn. On the way over, we talked of the changing face of the Black Mountains: the effects of acid rain, logging, and the landslide of 1976. While shoveling sawdust, I heard the family history of the sawyer and learned about the settling of the valley. After the sawdust was in the barn, my friend showed me his garden, orchard, and woodlot. He showed me his relationship with the land. Now I see the mountains with his eyes.

I listen to the world around me, too.

One of my favorite hiking trails has been an important voice in my education. Hiking the same trail dozens of times, in morning light and at night, trudging through snow or blackberry bushes, I have seen the many faces of the plants and creatures who live there. The subtle variations from day to day and the dramatic seasonal changes have inspired much poetry and song.

One tree along the trail has become a special friend. One blazing afternoon, wearing only shorts and a T-shirt, I was caught in a sudden, chilling rain. Fearing hypothermia, I stayed in the shelter of the hemlock, relatively warm and protected, waiting for the cloud to dry. Another day I climbed into the hemlock's embracing limbs to view the valley. I have eaten its crystallized sap and watered its roots. Through this deeper sharing I have come to understand the ecological niche of this tree; I have learned its story. (Please see "Bare Stone" on page 192.)

It's a story that begins with the birth of Gaia; migrates with the continents and glaciers; involves the wind and rain, mosses, fern, centipedes, roly-polies, and a mouse—a mouse who drops a seed into a moist clump of moss. From this seed the story of the hemlock sprouts.

I encourage you to take time to sit quietly and listen to the stories of your land.

Telling the Stories of the Land

Imagine yourself a water molecule that flows from the clouds through the ecosystem to your faucet and back to the clouds. What are the intervening adventures? Trace your lunch backward to its point of purchase, or the processor, to the soil where it grew. Or, trace it forward. Where does it go from your belly?

Mark your calendar on the day you see the first robin of spring, the first and last frost of the growing season, the day your first tomato ripens, and the day the first snow falls. Use the same calendar year after year to note the night that the katydids begin to sing, the day you get your first mosquito bite of the year, and the week of peak color on the maple trees. This folk wisdom can be woven into any number of stories. For example, one tale could be told from the perspective of a creature telling what it does (and what its friends and foes do) to adapt to the changes of color, temperature, and the available food supply.

I encourage you to tell these stories. Whisper the names of the wildflowers. Wade knee-deep in an icy creek, singing the songs of frogs. Imitate owls. Share the silences with the oaks and maples.

The land is full of legends clamoring to be heard! ❖

*The bristle cone pines
have been singing the same song
for 6,000 years.*

Acknowledgments

I would like to acknowledge all of the creatures of the Earth who have helped me along my journey and inspired me to write these tales. Most notably, I would like to thank the hemlock, dragonfly, mountain lion, milkweed seed, salmon, water drop, molecule of iron, owl, rattlesnake, spider, crow, John James Audubon, and the children of the Maine Healing Arts Festival. They told me their stories so I could tell them to you.

I would also like to acknowledge the giants upon whose shoulders I stand. There have been countless writers in the American naturalist tradition whose voices have helped me to find my own. Some of them include Henry David Thoreau, Ralph Waldo Emerson, Walt Whitman, Aldo Leopold, Rachel Carson, Gary Snyder, Mary Oliver, Barry Lopez, Annie Dillard, and David Rains Wallace. If you have not read their stories, visit your local bookstore or library today.

This book would not have been possible without the technical support of my office staff, to whom I am perpetually grateful. Thank you, Kathy Gess, Jane Menk, and Bev Loos, for typing various parts of the manuscript. Thank you Laurel and Kim, for retyping the manuscript for the new edition. I also want to thank Sue Blough, Sheila Price, Joy Connolly, and Diane Coon for all the work you have done that allowed me more time to work on stories. Thank you Gail and Tim from *Green Teacher* for editing various incarnations of a few articles that found their way into this new edition. A special thanks to James Thrush for your encouragement and support.

To the students and staff of the Seneca, Illinois, summer school program, who gave me a chance to field test many of these stories and lessons, I owe a debt of gratitude.

I would also like to thank Susan Zernial and Stacey Chisholm of Libraries Unlimited, as well as editor, Constance Hardesty, for turning my rough notes into a fine book. And thank you Cyndee Anderson and Rajalakshmi and the team at Apex CoVantage for shepherding LFL 2.0.

And a special thank-you to Vin Luong for his illumination of these stories!

And finally, I would like to thank my wife and twin daughters for listening to my stories, encouraging me, pushing me when I was tired, and for allowing me the space to finish the manuscript.

Introduction

There is a lot of talk about creative writing across the curriculum, the whole-language approach, and the art of teaching thinking skills and problem-solving strategies. *Learning from the Land* is a natural tool for accomplishing all these things, and for introducing ecological concepts and getting students excited about science as well.

I had two goals in mind when writing this book. The first was to provide students with the opportunities for adventure and for empathy with wild creatures; to bring scientific facts to life. The second was to teach basic science skills in a way that integrates creative writing and storytelling.

This book contains a dozen original stories that incorporate an array of scientific concepts. Because of the way the stories are structured, they can easily be adapted for listeners of any age or level of knowledge. Eight-year-old children have responded to these stories with questions that showed they were beginning to grasp concepts like plate tectonics and geological time. Yet with the right combination of unusual facts, poetic metaphor, and allusions to the complexity of nature, these same stories have held professional naturalists' rapt attention. There is something here for everyone.

Following each story are a few student-tested activities that cross curricular boundaries and inspire creative writing, problem solving, and a deeper understanding of nature. In addition, *within* each story are several ideas for quick, hands-on activities that can be easily adapted to various grade levels with more or less preteaching and step-by-step coaching. (See page 30 for more details on these activities.)

Obviously, this book is just a starting point. It provides stories and activities you can use with your students now. More important, these stories serve as models for students to use in writing their own stories. The ultimate goal is to have students move into the outdoors to observe ecological processes firsthand, measuring and evaluating what they observe, and then creating stories to tell their peers.

This book gives you just a few of many possibilities for storytelling and follow-up activities. When you are developing your own ideas, try to include as much direct contact with nature as feasible. Most schools are within a short drive of a state park, national forest, small creek or seashore. Closer to home, a neighborhood park, a tree behind the school, or even the classroom terrarium can provide contact with nature.

If you love a walk in the woods, if you hunt or fish, or if you enjoy the birds at your backyard feeder, you have access to a wealth of stories. You can translate the whispers from the fields into tellable tales, you can draw on your experience in the outdoors to inspire and educate your students. You can tell the stories in this book, and you can use these stories as models to create your own stories and experiential activities.

How the Stories Developed

All of the stories in this book were developed in the oral tradition. When I discover the kernel of an idea, I let it germinate in my imagination. I carefully water and fertilize it by discussing the idea with friends and colleagues. I bring it into the sunshine gradually by practicing on the nieces, nephews, and the children of friends. I allow it to mature and bear fruit through countless public performances at schools, nature centers, conferences, and festivals. Only after telling a story 100 times do I write it down. Often I record a live performance and transcribe the spoken word. I then edit the stories to meet the rigors of the printed page.

Although the stories are lightly edited, they are as much like the original spoken version as possible in order to capture the dynamics and electricity of a live performance.

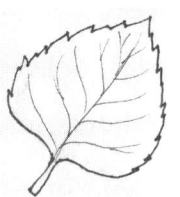

Storytelling, more than any other medium, has a special power for inspiring and educating listeners. For this reason I would like to encourage you to read the story a couple of times, and then put the book down and *tell* these stories.

Becoming a Teller of Tales

You are already a teller of tales. Every day you tell your own stories. You might tell them straight, or you might embellish a bit: add a few details here, a note of explanation there. Through these little embellishments, you make the story your own.

The stories in this book are my stories; they reflect my experiences, my local environment, my embellishments. They model for you all the details and embellishments that go into a story that is told, rather than read.

As any storyteller would do, feel free to change the details to fit your local environment and to reflect your experience and personality. Use the names of local rivers and mountains. Include plants and animals that are indigenous to your region. Change the names of characters to include your friends and relatives. Edit scenes that do not fit your personal history, and add scenes that do fit in with what you know. For example, "The Web" is about a fishing trip on one of my favorite lakes. If you have ever been fishing, you could easily change the details to suit your favorite lake or river. You could even change the species in the story from bullfrog to leopard frog, heron to egret, large-mouth bass to tarpon. Many of the stories were created as a framework: it's up to you to flesh them out with the details of your local environment as I often do when performing in different ecosystems.

How to Learn a Story

To tell a story well, you must enjoy it yourself. So, the most important thing in choosing a story to tell is to find one you really enjoy and can imagine yourself telling. If you don't love the story you tell, do you think your audience will?

After you have found a story, read it over several times. Read it once out loud for feeling, rhythm, and tone. Read it again to get to know the characters. Imagine you are each of the characters as you read. Read it again to get to know the place or setting. Imagine you are in that place, seeing all the details not in print. Read it one more time to learn the order of events. Stand up and read the story while mapping out body language and gestures.

If you read a story out loud seven times, then the words and images, the feelings and gestures will be imprinted on your psyche and you will know the story.

Do not memorize the story. I have often seen beginning tellers forget one word and then forget the rest of a story. Simply remember the important scenes, feelings, images, and phrases. (Sometimes I write a brief outline of the five main events to help me remember the sequential order of the story.) If you remember the bare bones, then you can flesh it out differently each time you tell it, adding details or skimming over things to suit the audience.

If you imagine that the story happened to you, then telling the story is like telling someone about your day. The words are less important than the images and feelings. Be in your story and let *your* words tell it.

Practice, Practice, Practice

The three most important things to remember when learning a story are to *practice*, *practice*, and *practice*. These are the three keys to success. Tell the story to every member of your family, every friend who will listen, the dog, the cat, and the goldfish. In the beginning, ask for positive feedback only: What are you doing right? What is working? Learn to screen the comments, paying attention to the helpful ones and disregarding the others. Only after you have learned to screen the feedback should you ask for constructive criticism. Remember, this is your story, so you must trust your judgment.

Telling the Tale

Before you begin telling a tale, relax. Pause for a moment of silence, take a deep breath, and let the story unfold. A little nervousness is natural and helps you to focus. You are nervous because you care about what you are doing and want to get it right. Reframe your nervousness in this light and see it as a positive force to be harnessed to help you do your best. Then you can have fun and enjoy the audience response. Remember, your listeners are rooting for you; they want you to succeed because it means they will have a good time, too!

Prelude

A good prelude grabs your listeners' attention, prepares them to listen, and eases (or jolts!) them into the tale. Begin with some history or background;

something about the author or source of the story: a song, poem, or chant: a question or purpose for the listener; or some intriguing sound effect.

To model this technique, each story in this book begins with a prelude that I use in my shows. Some of the preludes are personal. For example, the prelude to "The Web" is about my friend Lars and me. You will need to tailor the prelude to your audience and yourself.

Imagination

Imagination is the most important tool in a storyteller's bag of tricks. You must imagine the character to know what voice or posture to use. You must imagine the setting before you bring the audience into it. Imagination will allow you to embellish and personalize the story, tailoring it to the audience and making it uniquely yours. And, if you lose your place in the middle of a story, your imagination will allow you to make it up as you go along!

Use your imagination to live the story as you tell it. Whenever I am learning a new story, I always spend some time imagining each character, each scene, and each event. That way, when I tell the story, the words come easily. Instead of memorizing the words of someone else's story, I simply tell the story as I have experienced it in my imagination as if it happened to me yesterday.

Involve the Audience

The audience is your partner in the storytelling process. Involve your listeners with one or several of these techniques:

First and most important: Make eye contact! Simply looking into each person's eyes, speaking to them, is the best way to hook your listeners.

Ask rhetorical questions: Have you ever felt (seen, done) this before? Do you know someone like this? Rhetorical questions hook your listeners intellectually.

Use local metaphors. Compare elements of the story with tangible places, events, and people. Say, for example, "the boulder was as big as this room," or "The owl's wings were as long as my arms."

Use refrains. Many stories feature a chorus or a key line that repeats several times. Invite your listeners to repeat key lines with you like a chorus.

Encourage participation: give your listeners a job by assigning them to a character. For example, I will ask my listeners to be the wind (to make wind sounds) when I tell the story, "The Seed."

There are many ways to involve the audience emotionally, cognitively, physically, and imaginatively. Often this is not written into the story, but with a little imagination and rehearsal you can find ways to celebrate and build on the relationships between the teller, the tale, and the listener. Open your intuitive nature and tune into your audience. Watch carefully and listen attentively. Build on this invisible bond to bring your listeners under your spell!

Use All Five Senses

We perceive the world through all of our senses. In telling stories we help our listeners experience an imagined world. To make this happen, minister to all of the senses. Describe or allude to flavors. Remind your listeners of smells pleasant and profane. Specify the textures of clothing and skin, the textures of moods and milieus. Stimulate the senses with noises: clap, stomp, hoot, and squeak. Draw elaborate word pictures.

Voice

Speak in a voice that is loud enough to be heard and clear enough to be understood. Pace yourself and give your audience time to process what they hear.

Most beginning storytellers speak too quickly; it is best to let the words roll off your tongue at a pace that allows listeners to relish the sounds and the meanings of what is said . . . Pause . . .

Convey the feelings that are natural to the character and situations involved in the story. Avoid a melodramatic, syrupy storyteller's voice. Feelings are usually best understated rather than overdone.

Keep your telling lively, frequently changing tone, pace, and color. Accents can be very effective if used appropriately and accurately.

Sound Effects

Sound effects add humor and suspense. They can set the mood and grab the attention of a listener who starts to stray. In these stories there are many opportunities to hoot and howl, swoosh and growl. I have written in many sound effects, but feel free to add more or to change them.

Gestures and Body Language

Use gestures that come naturally to you. If you are the type of person whose hands move whenever you talk, then you will use more gestures than other people. If you do not talk with your hands, don't try to force it. Instead, find a style that suits you.

If you like, you can learn a few basic signs in American Sign Language. Sprinkle them through your story to add another layer to your telling.

Facial expressions are a great tool for expressing emotion. Sometimes I will pause in the middle of a story and give the audience *the look*. Without words, my listeners will know exactly what I am thinking (or I give them a moment to figure it out).

Walk the way you think the characters would walk. Stand the way you think they would stand. In these stories there are countless opportunities to stretch your arms in flight, stalk, or climb an imaginary tree.

When you are ready to try more complex modes of expressing the story through body language, try pantomime and dance. Workshops in

either art form will give you added appreciation for the subtlety and richness of nonverbal communication.

The End

Supply a neat and tidy ending to the story. There are many classic endings that can be adapted to almost any story, for example, "and now this story begins again", or " and to his day . . .". After your ending, pause for a moment of silence. Bow and graciously accept the applause!

Be Yourself

Each teller is different and tells stories differently. Some make great sound effects; others excel in movement and mime. Find a style that feels comfortable. Try new things, experiment. Discard what does not fit and practice what feels right. Remember, you are already a teller of tales!

Have Confidence

With the right amount of practice and a little faith in your ability, you too can captivate an audience. With *Learning from the Land* you can help your students understand their relationship with local ecology and celebrate the web of life.

A Word about Experiential Education and Hands-On Science

One of the goals of this book is to encourage teachers to provide hands-on opportunities for students to conduct scientific research, to replicate experiments and to interact with the material presented in each story.

Each story introduces several large concepts in the fields of natural history and ecology. These stories build a conceptual framework, but each story is also designed to build science skills, to train listeners to think like scientists and conduct experiments. The activities following the stories help students to think like a scientist *and* think like an artist by encouraging them to write creatively about their discoveries in science.

Conducting Experiments while Telling the Tale

Within several stories I have woven experiments or opportunities for students to formulate hypotheses. It is exciting to interact with students, allowing them a few moments to come up with a hypothesis, brainstorm possibilities, design an investigation, test their theory, and then evaluate their thinking.

"The Ballad of Rusty and Nancy" incorporates a mini-experiment to discern why an owl has one ear higher than the other. I usually do this by asking each student to choose a partner. Then I say, "Why do you think an owl has one ear higher than the other? What are the advantages? Remember, a hypothesis is more than an educated guess, it is taking what you know and figuring out a testable reason it might be. Think about what you already know about owls, and come up with a hypothesis." I give the students one minute to discuss the question with a partner, and then I bang on my drum to get their attention. Usually, I do not call on students at this point. Instead I say, "Let's test your theory." I choose 5 to 10 students as a control group and ask them to keep their eyes open. I ask the rest of the students to close their eyes. Then I move around the room clapping, sometimes holding my hands high, sometimes low. I ask the students with their eyes closed to point in the direction they hear the clapping.

The control group observes the students with their eyes closed to affirm which ones correctly point to where I am standing as I clap and move around the room.

Finally, I ask the student to open their eyes and discuss with their partner how they knew which direction the clapping was coming from, to collect and analyze the data. Does this support their hypothesis or not? At this point, I call on several students to answer either or both of the questions: How did you know where the sound was coming from? Why does an owl have one ear higher than the other?

I always emphasize that a partial answer is good and that several answers together give us the whole picture. This is how science works. No one knows all the answers, but if we put our heads together we can figure it out!

In similar (but simpler) fashion, I ask students to discuss why a great blue heron swallows its prey head first, (see "The Web") and how a rattlesnake became stuck under a rock (see "The Rattlesnake that Tamed a Boy"). When developing your own science stories, pause and answer this question: what are ways you could engage students in science process skills in the middle of your stories?

Posing Rhetorical Questions

Many times throughout each story I pose rhetorical questions to challenge my listeners to think like scientists. Many of these questions could be explored in more depth or used as springboards for research after the story is told. One of the primary goals of each and every story in this collection is to motivate hands-on discovery!

Hands-On Activities to Encourage Scientific Discoveries

Science is a verb, something you do. Following each story are several hands-on experiments that model various levels of inquiry. These activities allow students to interact with the concepts introduced in the story. Most of the activities can be done with few materials and minimal preparation. The most important factor is that students are given a chance to *do* science, not just hear or read about it. These activities facilitate easy, efficient opportunities for maximum involvement.

Finally a number of suggestions for activities are mentioned. Full details about how each activity is performed are unnecessary and would make this book too long. The ideas are offered as springboards for student lead inquiry. As environmental storytellers our goal is to inspire inquiry and facilitate student-led investigations.

Feel free to supplement the ideas presented with additional activities and experiments. Most importantly, allow students' questions and ideas to lead them to design investigations and follow their natural curiosity into genuine inquiry!

Writing Creatively about Science

Following several stories are worksheets to help students conduct research and write their own science story. Most of these sheets resulted from my classroom teaching experience, from my experience as the outreach educator for the Cincinnati Museum of Natural History, or from my more recent work as a consultant for various nature centers and science museums. Students may work on them individually, but for the most part, the worksheets are better used in small cooperative learning groups with extensive teacher supervision. In this way students can help each other with the research and peer-edit their writing. Many of these sheets are also available on my website as PDF files with hot links to the websites needed to conduct the research: http://www.foxtalesint.com/LessonPlans/LessonPlans.

Teachers and students could use these writing projects to build a portfolio, bind into books, post as a blog or create a file for future students. In this way, future research is assisted by prior projects.

Obviously, the activities, ideas, and worksheets are just the tip of the iceberg. Each time I reread the stories I see dozen of openings for experiments in each tale, countless opportunities for students to explore the natural world and acquire the skills they need to think like scientists.

A Note on Anthropomorphism and Folklore

Anthropomorphism means presenting or interpreting nonhuman entities in terms of human feelings and attributes (anthropo = human, morph = to change). Although fables and folklore make use of anthropomorphism, they also contain a great deal of information about how a culture relates to its environment. This is one of my favorite aspects of cultural biogeography, that is, how a culture reflects its ecosystem through its folklore. Anthropomorphism in traditional fables and folklore is not meant to imply that animals actually have human emotions; rather, the animals in fables are metaphors for humans.

Although most of the stories in this book are written with folklore models in mind, great pains were taken to avoid excessive anthropomorphism. Simply naming a molecule of iron oxide Rusty obviously violates this principle. But a careful reading of the text reveals that efforts were made to present the animals with few human feelings or traits. Some personification is helpful in developing empathy and building a bridge of understanding. When used carefully, it adds to the story and its telling. Nevertheless, all of the stories are firmly grounded in science fact, not science fiction. It is this balance between fact and techniques for building empathy that make these learning *stories*.

All of this raises the murkier question of anthropocentrism. Are humans the most evolved life form? Western science tends to regard humans highly, often disregarding the wisdom of other cultures that see humans as merely another strand in the great web of life. One of the goals of this book is to create stories that value the inherent wisdom of nature. Again, this requires a broader approach than the traditional just-the-facts model. Although they use scientific models for investigation, most scientists agree that such models cannot answer all questions. Complementing the traditional approach is much recent research that explores the ways plants and animals communicate and think or feel. Although it is difficult to design studies that distinguish instinct from intellect, more and more scientists are agreeing that animals do use reason, are good problem solvers, and have profound feelings.

Admittedly, the stories in this book raise a number of these questions without fully answering them. The ideas of anthropomorphism and anthropocentrism provide rich possibilities for classroom discussion. For example, recent research shows that trees release chemical signals when they are attacked by disease; these signals warn other trees, who in turn produce thicker bark as protection. Parrots and dolphins can answer questions posed by their trainers. Are these examples of communication? How do they differ from human communication?

Creative Nonfiction versus Anthropomorphism in Student Stories

When students write science stories based on their research, they tend to anthropomorphize the animals. The challenge for students is to write science facts in story form, creative nonfiction, without extensive anthropomorphism.

Encourage students to analyze the facts (what they can observe and measure) and write stories based on those facts. Explain that the goal is to see what we can learn from studying the natural world and to write stories that celebrate our roles and relationships within the complex web of life. While encouraging creativity and self-expression, suggest that students make every effort to remain grounded in the realm of science.

Have students explore issues of anthropomorphism and anthropocentrism in editing and rewriting their stories. For example, instead of having animals speak in human languages, the students could explore the animals' calls, signals, body language, and pheromones. Simply deleting anthropomorphism can tighten up a story and improve the flow. Challenge students to identify murky or questionable aspects of their story and conduct additional research to increase the scientific accuracy of their tales.

Students as Storytellers

Encourage your students to tell their tales as a tool for helping them to rewrite and edit them. After the students have told their stories to their class, give them a chance to tell their stories to other classes and grade levels. I have seen it again and again: If students tell their stories—even if it is only to a partner—their writing becomes much more fluid, detailed, and cohesive. *Oral language development is an important step in improved reading and writing.*

Whenever you ask students to perform for a large group, give them ample opportunity to rehearse in smaller, supportive groups. Here's a way to structure this:

1. The student practices at home. (You might ask a parent or older sibling to write a note with positive feedback for the student. Have the student show you the note to verify that she did practice at home. Offer extra credit for every listener's signature.)

2. The student practices with a partner.

3. Four students in a cooperative group tell their stories to each other.

4. Each student performs for the entire class.

Throughout this process emphasize the positive. Give students opportunities to compliment each other, and make it clear that no criticism is allowed. Tell students, "Tell your partner one thing you liked about the way he or she told the story," or "Raise your hand if you have a compliment and the performer will call on you." Some students need practice being nice; be sure to model positive feedback for them.

Teaching Basic Science Skills

Science is a process, a way of thinking, a way of interacting with the world. This process requires a basic set of skills. Figure 1 (page xxv) lists the basic science process skills as determined by Funk, Fiel, Okey, Jaus, and Sprague (*Learning Science Process Skills*, Dubuque, Iowa: Kendall/Hunt, 1979.) Figure 2 (page xxvi) is a form to help you evaluate the ways in which you are already encouraging students to practice these skills and ways you might encourage such practice in the future.

Conclusion

Science is about discovery, about exploration, and about communicating our discoveries to others. More and more, it is recognized as a creative effort, with leaps of imagination based on sound observation and logic. Science 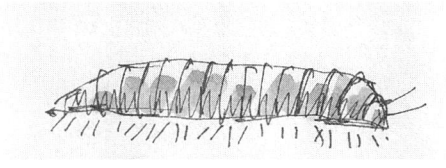 and storytelling are a natural pair, for science is the story of our world. Help your students learn to listen to the songs of the cricket, and soon they will blossom with songs of their own!

If a child is to keep alive his or her sense of wonder . . . he or she needs the companionship of at least one adult who can share it, rediscovering the joy, excitement, and mystery of the world we live in.

(Rachel Carson, from *A Sense of Wonder*)

Science Process Skills

Basic Science Skills

Observation
    To look, listen, touch, taste, and feel; to use the five senses to acquire detailed specific information.

Metric Measurement
    To measure height, weight, length, temperature, and time using the international standard.

Classification
    To organize or impose order on a collection of objects, events, and living things, observing similarities, differences, and interrelationships and grouping them according to an arbitrary but useful purpose.

Communication
    To clearly define ideas, directions, and descriptions for others to comprehend. (One use of communication is to document research so that others can repeat the study to ascertain its validity.)

Prediction
    To forecast a future observation based on careful observation and inferences about relationships and observed events.

Inference
    To recognize patterns and expect these patterns to recur under the same conditions.

Integrated Science Process Skills

Identify Variables
    To identify anything that is likely to change or vary during the course of an investigation. There are two kinds of variables: manipulated variables and responding variables.

Formulate Hypotheses
    To construct an educated guess based on the relationships among variables.

Design Investigations
    To set up a planned situation to yield data about the accuracy of a hypothesis.

Reorder, Analyze, and Draw Conclusions
    To collect data from an experiment, organize it in a meaningful way through graphs and exposition, and use it to support a hypothesis.

Figure 1. Basic science skills. Adapted from *Learning Science Process Skills* by Funk, Fiel, Okey, Jaus, and Sprague (Kendall/Hunt, 1979).

Teaching Basic Science Process Skills

More than a body of information, science is really a way of seeing and thinking, a way of measuring and understanding our world. Science is a verb, something you do. Following is a list of basic science process skills that will help your students think as scientists. Below each skill statement are three blank lines. On each line write an activity that you do use or that you may use in your classroom to teach or apply the skill.

Basic Skills

Observation

To look, listen, touch, taste, and feel; to use the five senses to acquire detailed specific information. (Example: Write a descriptive paragraph about an apple. Use all five senses.)

_____

_____

_____

Metric Measurement

To measure height, weight, length, temperature, and time using the international standard. (Example: Use metric measurements to measure the height, length, width, and weight of a favorite book.)

_____

_____

_____

Classification

To organize or impose order on a collection of objects, events, and living things by observing similarities, differences, and interrelationships and by grouping them according to an arbitrary but useful purpose. (Example: Cut pictures of animals from magazines or calendars. Group the pictures into categories the students define or create. Discuss the groupings or categories and compare them to the established categories of kingdom, phylum, family, genus, and species. Or, have students invent a classification system for winter coats. Have students work individually or in small groups to create a classification system, then discuss and compare the various classification systems created.)

_____

_____

_____

Communication

To clearly define ideas, directions, and descriptions for others to clearly comprehend. One use of communication is to document research so that others can repeat the study to ascertain its validity. (Example: Write step-by-step instructions for making a sandwich.)

_____

_____

_____

Prediction

To forecast a future observation based on careful observation and inferences about relationships and observed events. (Example: Each day for one week note what time buses arrive at the school [at the end of the day]. Then predict which bus will arrive first on the next five days.)

_____

_____

_____

Inference

To recognize patterns and expect these patterns to recur under the same conditions. (Example: Working in small teams, diagram the floor plans and desk arrangements in several classrooms. Then predict how other classrooms might be arranged based on grade level, relationship to the hall, and other variables.)

_____

_____

_____

Integrated Science Process Skills

Identify Variables

To identify anything that is likely to change or vary during the course of an investigation. There are two kinds of variables: manipulated variables and responding variables. (Example: Brainstorm and discuss *what-if* questions. If this changed, what else would change? For example, if our classroom were 10 degrees colder [manipulated variable], what other things would change, and how [responding variable]?)

_____

_____

Formulate Hypotheses

To construct an educated guess based on the relationships among variables. (Example: Brainstorm a list of *why* or *how does that work* questions: On a clock, why does the minute hand move faster than the hour hand? How does that work? Discuss possible answers based on what you know and what you guess to be true.)

_____

_____

_____

Designing Investigations

To set up a planned situation to yield data about the accuracy of a hypothesis. (Example: In answer to the question about how and why the hands on a clock move, write a one-page plan of action to answer the questions. The goal is to answer the question, How could we find out . . .? For example, you could carefully take apart a broken watch and compare the size and shape and position of the gears attached to the various hands.)

_____

_____

_____

Reorder, Analyze, and Draw Conclusions

To collect data from an experiment, organize it in a meaningful way through graphs and exposition, and use it to support a hypothesis. (Example: After completing the investigation, present the findings orally and in writing. For example, draw charts of the gears in a clock and explain the relationships among them.)

_____

_____

_____

Here are a few books every naturalist should know by heart:

Leopold, Aldo. *A Sand County Almanac*. Ballantine Books, 1970. 13: 978-019514 6172. Leopold is the father of the modern conservation movement and this book is the bible of that movement. Several of the stories in this book are inspired by his essays. His keen insight and genius ripples off of every page in the way he blends sharp scientific analysis with poetic metaphor and descriptive flourishes. At once philosophical and practical, his observations of an ecological year on his Sand County farm have implications for the fate of the Earth. Several chapters are also adaptable for the telling.

Watts, May Theilgaard. *Reading the Landscape of America*. Nature Study Guild Publishers, 1975. 13: 978-0912550237. Out of print for a long time, finally available again, this is a classic work and in all honesty, one of the key inspirations for my life work! She brilliantly trains the average reader to be a nature detective and read the ecological history of the landscape by giving dozens of examples of her hikes, train travel, and even an airplane adventure where she reads the big picture at 10,000 feet! A walk in the woods or prairies or mountains will never be the same.

Stilgoe, John R. *Outside Lies Magic*. Walker, 1998. 10: 0802775632. Few books have challenge the way I think and look at the world like this one has. Stilgoe trains the eye and mind to see beyond the surface of urban landscapes and read the history of the place in the same way May Theilgaard Watts opens our eyes to the layers of natural history in her book. A walk in your neighborhood will never be the same.

Carson, Rachel. *Silent Spring*. Mariner Books; Anniversary edition, 2002. 13: 978-0618249060. A gifted writer who could synthesize reams of scientific data and draw astounding conclusions that actually changed the way we think about the wild world and our relationship to it. This book literally wrote laws about chemical use around the globe and the earth is healthier for it. Few folks know or remember she won a Pulitzer Prize before writing *Silent Spring*. If you read nothing else, read the introduction, "A Fable for the Future." Learn it and perform it!

Wallace, David Rains. *Klamath Knot*. Sierra Club Books, 1983. 13: 978-0871568175. While hiking in the Cascades I read this book and it opened up layers of natural history, deepening the experience in a way few other books could. A near-perfect example and high-minded model of the kind of writing I hope these lessons inspire.

Louv, Richard. *Last Child in the Woods*. Algonquin Books of Chapel Hill, 2008. 13:978-1-56512-605-3. Louv makes startling claims about the importance of spending time outdoors in order for children and adults to be more fully human, and he provides the research data to support these claims! Spending time in the wild is key to brain development as well as sanity, it is vital for physical development, and as an antidote to our obesity crisis, it stimulates creativity and develops problem-solving skills. In the updated and expanded version of this best-selling book, he goes several steps further in providing antidotes to what is now being called Nature Deficit Disorder.

Suzuki, David. *The Legacy*. Greystone Books, 2010. 978-1-55365-570-1. If *Silent Spring* launched the ecology movement, then *The Legacy* offers the wisdom to see us through to fruition. Though it has a dark and foreboding beginning as he rattles through the daunting statistics of our ecological peril, he offers great hope, inspiration, and practical advice for ways to survive climate change and move toward a sustainable future.

And here are few other collections of stories and lesson plans that are highly recommended:

MacDonald, Margaret R. *Earth Care: World Folktales to Talk About*. Linnet Books, 1999, 208024263. These thought-provoking folktales about the larger issues of responsibility and consequences also have an environmental ethic at their core.

Livo, Lauren. *Of Bugs and Beasts; Fact, Folklore, and Activities*. Teacher Idea Press, 1995, 13: 978-1563081798. A fun collection of stories and even better collection of lesson plans.

DeSpain, Pleasant. *Eleven Nature Tales*. August House, 1996. 13: 978-0874834581. These are very tellable folktales with ecological implications.

Gail, Judy and Linda A. Houlding. *Day of the Moon Shadow. Tales with Ancient Answers to Scientific Questions*. Libraries Unlimited, 1995. 13: 978-1563083488. If you believe, as I do, that other cultures use folktales to explain natural phenomenon, then this is a very useful collection of stories and lessons.

Caduto, Michael J. and Joseph Bruchac. *Keepers of the Earth*. Fulcrum, 1991. 13: 978-1555913854.

Caduto, Michael J. and Joseph Bruchac. *Keepers of the Animals*. Fulcrum, 1992. 13: 978-1555913861.

Caduto, Michael J. and Joseph Bruchac. *Keepers of Life*. Fulcrum Publishing, 1994, 1-55591-186-2.

Caduto, Michael J. and Joseph Bruchac. *Keepers of the Night*. Fulcrum, 1994. 13: 978-1555911775. This now-classic collection of Native American folklore and easy-to-follow lesson plans are required reading for every environmental educator.

Seed, John, et al. *Thinking Like A Mountain*. New Society Publishers, 1988. 13: 978-0946097265. Unlike most collections of lesson plans, this book is an integrated, intensive workshop that helps participants experience deep ecology. Best organized as a day-long event, "The Council of All Beings" presents a non-threatening avenue into shamanism, and an attempt to learn from the wisdom of other creatures who share our home. I have presented similar workshops for adults and kids in formal and informal settings and would highly encourage you to create a council for your community!

Shaffer, Carolyn and Fielder, Erica. *City Safaris*. Sierra Club Books, 1987. 13: 978-0871567208. This is a wonderful collection of field trips designed for urban environments.

Cornell, Joseph. *Sharing Nature with Children*. Dawn Publications, 1998. 13: 978-1883220730. This classic work on environmental education is filled with fun games, many that work inside but are much more fun outdoors! If you purchase just one book of ecological lesson plans this year, this is it!

And my favorite magazine for environmental education: *Green Teacher*, which I highly recommend to any educators, formal or informal, who wish to expand their knowledge of the environment and find classroom-tested ideas for teaching ecological issues. http://www.greenteacher.com/

The Ballad of Rusty and Nancy: A Trilogy of Creation and the Mineral Cycles

A single protein molecule or a single finger print, a single syllable on the radio or a single idea of yours, implies the whole historical reach of stellar and organic evolution. It is enough to make you tingle all the time.

—John Platt, from *The Steps to Man*

Comments to the Teacher

WHEN I TAUGHT SIXTH-GRADE SCIENCE, I used this story at the beginning of the year to introduce the conceptual framework for the entire year. Throughout the year I made references to this story, so the pieces of the science curriculum fit into a larger picture. This is the reason I begin the book with this story. You will also note that some of the other stories in this book borrow pieces of this story, expanding on them or using them with a different emphasis.

"The Ballad of Rusty and Nancy" describes the repeated appearance of two elements, iron and nitrogen. You can adapt this story to various topics. For example, you could focus on iron and nitrogen in the plants and animals of a specific ecosystem, like the rain forest or desert. Or you could focus on something more specific, like human anatomy and physiology using a story about Rusty the Red Blood Cell and the body's reaction to crisis from the internal point of view of one blood cell. You could focus on geology, including plate tectonics, geological time, the mineral cycle (rock to sand to sediment), and types of rocks (igneous, metamorphic, and sedimentary). Or you can keep it broad and focus on various branches of science. When I tell this story to teachers in an in-service workshop I challenge them to brainstorm a list of all the branches of science. I then challenge myself to follow Rusty through all of the branches of science the teachers have named. Everywhere that iron or nitrogen appear, Rusty and Nancy do, too.

It is easy to adapt this story to your local geological and environmental history. The version in this book is specific to the upper Midwest, where I live. When I tell this story in the desert Southwest or in northern Europe, I adapt it to the geological and environmental history of those regions. Although I change the names of the plants and animals, the concepts remain the same. For example, in the desert I focus on shallow oceans, sedimentary rocks, and ancient sandstone. Rusty moves through dinosaurs, the Anasazi, the Zuni, cacti, bats as pollinators, rattlesnakes, lizards, mule deer, and mountain lions. With Nancy I may explore strip mining, smokestacks with large emissions of sulfur and nitrous acid, and acid rain. Before telling this story, you may want to do a little research to adapt the story to your region. Read the story once before you visit the library, and simply plug local details into this flexible framework.

If this long story seems overwhelming, it is very easily shortened. When I perform it at a conference or festival it takes 45–60 minutes, When I told it to my sixth-grade class, I told it over the course of the three days starting with background information and the big bang on day one, Rusty on day two, and Nancy with follow-up discussions on day three. Because of time constraints I am often forced to clip off the big bang, skim the geology, and cut one loop of the food web to tell a 15-minute version of Rusty. To trim Nancy to a 5-minute version, skim the geology and petrochemical refinement, fast-forward her shipment on the boat, cut the details of the birthday party, and move right to the pollution. (All of this will make more sense after you read the story.)

Of course, I recommend you tell the entire story (and I have seen young children listen attentively to the whole thing), but in today's classrooms, there may not be time for the entire tale.

For example, I recently told a very short version of this story at the New Mexico Museum of Natural History, as part of hike through the museum for a student writing workshop. They have a beautiful wall that mimics a rock cut through the geologic layers of New Mexico. Standing on the stairway with a group of about 50 students, we covered a few hundred million years of local geology in less than five minutes using each strata of rock for a paragraph of material.

Also be aware that this story contains some graphic material geared for older students, including information about reproduction and some gory predator/prey scenes. This information is included because it reflects real life in the wild world, but it may be unsuitable for younger students. It is imperative that you read the story first, and adapt it to suit your audience.

And now the story . . .

The Ballad of Rusty and Nancy

This tale is actually a trilogy of shorter stories. It's a story that began 13.7 billion years ago and a story that is happening right now. You are a part of this story. The story is happening inside of you and all around you.

As I said, it all began 13.7 billion years ago. Well, not exactly. Scientists actually estimate that it all started about 12 to 14 billion years ago. But does anybody really know? Were you there when the universe was made? Was anybody there?

Think about it. In a way, you *were* there. The calcium in your bones, the iron in your blood, these are minerals. And the electrons and protons, the tiny atomic particles that later made these minerals, *were* present at the dawn of creation. These minerals are inside you today. So the next time somebody asks you how old you are, you can say, "Oh, at last estimate, I'm about 13.7 billion years old." When they laugh, you can tell them what I have just told you. It's true. The theory that most modern scientists believe goes something like this.

The Big Bang

Think about an atom, a cloud of electrons whirling around a tiny nucleus of protons and neutrons. For a long time, an atom was one of the smallest things known to modern science. Except now we know atoms are made up of electrons and neutrons and protons, and neutrons are made from smaller things called quarks. But atoms are so small that a million of them will fit on the tip of your fingernail. They are the building blocks of the universe. Maybe you have seen a symbol of an atom with the nucleus encircled by tiny electrons. I say symbol because atoms are so tiny no one has actually seen one. Based on research and theories, we can only hypothesize about their shape and size.

Think about our solar system, planets and moons whirling around the sun.

Add the comets that come into our solar system at odd angles, and the solar system looks a lot like an atom.

Our solar system is a tiny part of a larger congregation of stars called the Milky Way galaxy. Picture this: a cluster of stars at the middle of the galaxy, with smaller stars whirling around the outside edge. Our sun is just one of billions of stars, and our solar system is just one of the thousands of solar systems in the Milky Way galaxy.

Compared to an atom or a solar system, the Milky Way galaxy is enormous, humongous, gigantic . . . well, it's bigger than one can imagine. Yet all three of them—the atom, our solar system, and the Milky Way—have the same basic shape, a cluster in the middle with particles whirling around the center. Isn't that amazing? Keep these images in mind while I tell you this story.

Now, I want you to imagine everything in the room around you compressed together, squeezed down as tight as it will go. Take out all the space between things. Take out even the tiniest space that separates you and me and the chair and the ceiling. Take out the most important space, the spaces between the electrons, protons, and neutrons. If you could take out all the space between these things, then you could take this  entire room and compress it, squeeeeeze it down into something smaller than a marble. Of course that marble would weigh as much as this room.

Imagine taking out the space between the walls, between the buildings, between the houses and trees. Take out even the tiniest space between the electrons, protons, and neutrons. If you could take out all the space between these things, you could take your entire city and compress it, squeeeeze it down into something smaller than a softball. Of course that softball would weigh as much as your city.

Imagine taking out all of the space between New York and Los Angeles, between Peking and Paris, take out all of the space between oceans and mountains, rivers and forests. And don't forget, take out the smallest spaces between protons, electrons, and neutrons, and you could condense the earth, squeeeeze it down into something smaller than a basketball. Of course that basketball would weigh as much as the earth.

Imagine taking out all the space between the Earth, the planets, the sun, and the moon. Take out even the tiniest space between the electrons, protons, and neutrons. If you could take out all the space between these things, you could take our solar system and compress it, squeeeeeze it down into something smaller than a beach ball. Of course that beach ball would weigh as much as our solar system.

There are things in deep space that are compressed just like this. They are neutron stars, or black holes. Black holes are the remnants of supernovas, stars that have exploded. Black holes have so much gravitational pull even light cannot escape them.

Now imagine taking out all of the space between the solar systems and galaxies. Take out all of the space in between all of the matter in the universe. Take out even the tiniest space between the electrons, protons, neutrons. If you could take out all of the space between these things, you could take the entire cosmos and compress it, squeeeeze it down into something very small. How small? Well, no one knows for sure. They're still working on that.

Scientists believe that this is how the universe began, as one huge clump of matter. Where did the matter come from in the first place? No one knows for sure. But that's another story. Then, all those electrons, protons, and neutrons, all of that mass, all of that energy, began to undulate. All that energy began to grow, began to vibrate, began to expand and [*make loud exploding noises!*] *KABOOM!* All that stuff exploded!

Well maybe not one big bang, but the evidence says the universe was once a lot more dense and a lot hotter ... now it is expanding and cooling.

To give you an idea of how hot this universe was, some of this matter cooled to form stars. Some of the stars, like furnaces, heated molecules and fused them together to form larger molecules. Gravity drew clumps of matter together. Some clumps cooled even more to form liquids and solids—and planets. This is the Big Bang theory of the creation of the universe. Have you heard of this theory? Scientists came up with the Big Bang theory, in part, because they have huge satellite dishes, bigger than a football field, and these satellite dishes listen to the sounds in deep space. This is the sound they hear: k-k-k-k-k. Scientists believe that the k-k-k-k they are hearing is the echo of the big bang. No one knows for sure if they're right, but that's the theory.

So that's the Big Bang theory: the universe began as one ball of energy, and it exploded out into a whirling and twirling universe. That whirling and twirling and expanding is still going on today.

From that big bang, the Milky Way galaxy was formed. The Milky Way is a spiral galaxy that is whirling through space. You know what a spiral is, right? It's like a twisting thread. [*Make a spiral motion with hand.*] Many stars formed out on the edge of the Milky Way. Our sun is just one of those stars. Whirling around some stars are planets; this makes a solar system. The Earth—and Mars and Jupiter and several others—are the planets in our solar system. Some planets have 5 or 13 moons. Earth got only one—was that a fair deal?

Now think of this: the big bang set off the whirling, swirling universe. The Milky Way whirls through space. Solar systems whirl through the Milky Way. Planets whirl around stars. Moons whirl around the planets. And electrons whirl around the nucleus of the atom. In all that whirling and swirling are two very good friends of mine, Rusty and Nancy. I'd like to tell you their stories.

Rusty is a molecule of ferrous oxide. That's iron and oxygen, or rust. You've seen it. The rust on your bike, the rust on a car, the red in your blood is ferrous oxide. This is his story.

Rusty was there at the creation of the Earth, four and a half billion years ago. He was there when the molten magma cooled, when the rains began to fall and harden the crust of the Earth. He was there in all of that whirling and twirling—and he still is here today.

Let's stop for a moment now to think about geological time. Spread out your arms for just a minute. Reach out as far as you can. Look over at your right hand. That is the beginning of the Earth. Now look over at your left hand. That is the present moment, right now. From your right hand to your left hand, that is the span of time that has passed since the Earth was formed. That's four and half billion years.

Guess when dinosaurs appeared? Dinosaurs showed up past your left elbow, near your wrist. That was 250 million years ago.

Guess when humans appeared? If your right hand is the beginning of Earth and your left hand is the present moment, humans appeared on the tip of the fingers of your left hand. That's correct, your left hand! If you clipped your fingernails you cut humans off the timeline! Homo sapiens showed up not very long ago in geological terms: 25,000 years ago.

Let's get back to Rusty. He was here through all of this. Rusty was inside the crust of the Earth. He began as part of a large supercontinent called Pangaea, which later broke into Laurasia and Gondwanaland. What's a supercontinent? Well, at one point in time the continents we have today were all one piece. That's right—there was just one big continent, a supercontinent! If you could take a map of the world and cut out all of the continents—Africa, Antarctica, Asia, Australia, Europe, North America, South America—you would find they fit together like a puzzle. That's because, at one point in geological time, the curve of Africa fit right in between North America and South America. Can you imagine that?

How do I know? I know because rocks we find in New England are the same rocks we find in old England.

Slowly the continents drifted. Try to imagine that, if you can. Rub your hands together. Faster, harder! Push your hands together while you rub them faster, harder, harder, faster! Do you feel that heat? Do you feel that friction? Now could you imagine taking a rock the size of North America and rubbing it against a rock the size of Africa or Europe? Imagine all the heat that would make! All that heat would melt stone. That's just what happened as the continents drifted. And, don't forget, they're still drifting today. Hold on! North America is moving about an inch or so each year.

Rusty was part of that crust of the Earth near what is now Canada about 100,000 years ago, give or take 1,000 years. About that time, the Earth grew cold and snow fell faster than it could melt. As the snow piled up, it formed glaciers. Four times in the recent glacial epoch huge sheets of ice pushed out of Hudson Bay and Northern Canada.

Can you imagine a sheet of ice nearly a mile thick and hundreds of miles wide? Can you imagine a bulldozer that big, a bulldozer with a blade a mile thick and 100 miles wide? Everything in the way of that bulldozer would be plowed right over. All of the trees, all of the rocks—rocks as big as your house—would be pushed ahead of this bulldozer, this sheet of ice. As the glaciers moved down across North America, they bulldozed everything in their way. That's why the northern parts of Ohio, Indiana, and Illinois are flat as a pancake. How do I know? I know because Rusty was there. Rocks from northern Canada litter the fields of the Midwestern United States.

How did the rocks from Canada end up in Illinois? Some of the rocks that were bulldozed tumbled underneath the glacier. As the glacier moved on, the rocks rolled along underneath the ice. The weight of the glacier tumbled the rocks against each other, and that made the rocks split and crack. Sometimes chunks of rocks broke off and were left behind. If you live in the northern plains states, you may have found these chunks of granite, stones that look out of place. They are called glacial erratics, and they *are* out of place: they once belonged in northern Canada.

Finally the glaciers started to melt. To give you an idea of how big the glaciers were, think of them as giant ice cubes. One ice cube melted to make Lake Erie. Another ice cube melted to make Lake Michigan. The Great Lakes are remnants of glaciers.

Rusty was there. Rusty was in one of the rocks left behind by the glaciers. As the ice melted, the area warmed up. Fifteen thousand years ago, the weather in the northern plains was like the weather in Alaska today. How do I know? Well, if you send a probe into the bottom of one of the Great Lakes and pull up layers of mud and sediment, in the deeper layers you will find pollen from tundra plants, sedge, and willow. These are plants that grow in cold places—like Alaska. If you date that pollen to find out how old it is, you will find it is 15,000 years old. That's how I know that 15,000 years ago, Illinois had a climate like Alaska has today. Twenty thousand years ago, mastodons and woolly mammoths roamed. [Teacher: Remember, parts of this story are specific to the Midwestern United States. Feel free to change the details—such as the kinds of plants and animal life described—to suit your region.] Mastodons were enormous hairy elephants with huge tusks. And what was here 17,000 to 18,000 years ago? Saber-toothed cats, beavers bigger than you are, and dire wolves bigger than the modern wolves. They all lived here. How do I know? They left their bones behind. Today, farmers are still finding these huge bones in creek bottoms and along the edges of fields.

Many of those animals became extinct when the ice melted, but some migrated north and others adapted to changing conditions. The musk ox is one animal that migrated. It is still alive and well in the barren, northernmost regions of North America. The large beaver became extinct; smaller beavers adapted to the warmer climate.

The animals weren't the only thing that moved when the climate warmed up. If you were in Illinois not long ago—just 10,000 years—you would be in a deciduous forest. (Deciduous trees drop their leaves in the fall.) The oaks and hickories that grew in Missouri during the glacial period spread north as the area warmed up.

There was one huge oak tree, an oak tree so big around, three people holding hands in a big circle could barely reach around it. Can you imagine an oak tree bigger around than this and over 100 feet tall? Can you imagine? And of course, where there were forests there were bears and cougars, wildcats and elk.

So where was Rusty then? He was still in the rock under layers of soil.

About 10,000 years ago, that huge oak tree grew. It reached its roots down into the ground and it found the rock where Rusty was. That root found a crack in the rock. That root grew into that crack and that root eventually broke the rock in two. The root was like a straw. Slurp! It sucked the minerals out of that rock—Rusty included. Rusty went up the root, through the trunk of the tree into a branch, from the branch into a limb, from the limb into a twig, and from the twig into a leaf.

That leaf put Rusty to work. Leaves are like food factories for trees; they take sunlight and turn it into food. This process—turning sunlight into sugar and starch to feed the tree—is called photosynthesis. Can you say photosynthesis? That was Rusty's job: all summer long, he worked with chemicals and minerals in the leaf to make food for the tree. He worked every day that the sun was shining. If it was a cloudy day, he got to rest. Do you know why?

Rusty worked all day and slept all night. He worked all summer long. Then the ice and snow returned. It wasn't another glacier, no, it was winter. As frost settled in, the factory in the leaves closed down. The tree sucked most of the important minerals out of the leaves and stored them in its trunk and roots.

Yes, trees save and reuse those vital minerals. Recycling is nothing new! Mother Nature has been doing it for billions of years. Every fall, trees draw many of the minerals out of their leaves and store those minerals in the trunk and down in the roots for the winter.

So Rusty was drawn out of the leaf and stored down in the root as the snow fell. All winter long he waited. In the spring, when the sap rose and the little buds on the tree burst open, Rusty was drawn up from the roots, up through the trunk, but this time instead of going into a leaf, he became part of the flower.

An oak flower is a lot like a human. Some flowers are male, and some flowers are female. Most flowers, like tulips and dandelions, have the male and the female parts together. (This is an important part of science!)

Rusty went into one of the male flowers and he became a tiny grain of pollen. Pollen is like sperm made by a tree. Most of that pollen just blows out onto the ground. Maybe you have noticed in the spring that in the morning the cars and sidewalks are dusted with yellow powder. That's pollen. Billions and billions of grains of pollen are made every spring, and most of them are just wasted on the ground. But not Rusty. He found a female flower. And you know what happens when male and female oak flowers get together? They make acorns.

Rusty became an acorn. Acorns are made in the spring. Then, all summer long as the factory in the leaf makes food, some of that food is stored in the acorn. Rusty the acorn got fatter and fatter, plumper and plumper, all summer long. By fall, he was a big, fat acorn. Along came a squirrel. This squirrel was hungry. This squirrel was running around gathering acorns and hickory nuts and walnuts. This squirrel climbed up the oak tree, went out on a branch, and found Rusty. The squirrel ate Rusty the acorn! The squirrel digested the acorn, and so Rusty became part of the squirrel.

As a squirrel Rusty could climb tress. He could run out on the branches and he could jump from one tree into another. He could run down trees and around their trunks. He could gather acorns and store them in his cheeks. He could bury the acorns underground for the winter.

How many of you think squirrels remember where they bury all of their acorns? If you think they remember where they bury *all* of the acorns, raise your hand. If you think squirrels remember where they bury a few acorns, raise your hand. If you think squirrels don't remember where they bury any of their acorns, raise your hand. Surprise! Squirrels don't remember where they bury a single acorn. No. They just bury the nuts helter-skelter all over the place because nuts store better underground. Well, Rusty the squirrel buried acorns all through the autumn. Then, when winter came he went to sleep.

Squirrels do not hibernate like a bear or frog. They may sleep for a couple of days. When it's really cold they might sleep for six or seven days. Can you imagine sleeping for six or seven days? Imagine what dreams you would have. But if you slept for six or seven days, when you woke up, do you think you'd be hungry? Why, yes.

So, after a week, Rusty the squirrel woke up. He was very hungry. Now, if he couldn't remember where he buried all of those acorns, how do you think he found them? You're right, he smelled them. A squirrel's nose is so sensitive it can smell through a couple inches of snow, a couple inches of leaves, and a couple inches of dirt—up to one foot of whatever is covering those nuts!

Sniff, sniff, sniff . . . no nuts over there. Sniff, sniff, sniff . . . no nuts over here. Sniff, sniff, sniff . . . here are some nuts! Rusty the squirrel started digging down and found a nice plump hickory nut. He was so hungry he was not paying attention to what was happening around him. Rusty wasn't the only one who has hungry.

In a nearby tree was a great horned owl. That owl saw that squirrel. It lifted its large wings—wings longer than my arms, more than five feet long—and it soared down through the forest, dodging among the tress. Owls have special feathers on the front of their wings that slice the air so you cannot hear them fly.

Rusty did not hear the owl. Rusty the squirrel was busy eating a hickory nut when *snatch*! From out of nowhere came the owl. It grabbed the squirrel with razor-sharp talons. It tore at the squirrel's throat with a razor-sharp beak, and the squirrel's blood dripped onto the ground. With sharp talons the owl snatched up that squirrel and carried it back to the tree. As it tore chunks of raw meat from the bones of the squirrel, Rusty became part of the owl.

If you think being a squirrel was a lot of fun, you should try being an owl. Rusty the owl could fly though the forest at night. Can you imagine flying? She soared through the trees, even though her wings were more than five feet long. She dove left and right. Sometimes she had to stretch her wings straight up and down to move between the huge trees of the forest.

Owls are excellent hunters. On an average night, a great horned owl may catch and eat 15–30 mice. Would you rather have one owl in your neighborhood or a whole lot of mice?

We all know owls have big eyes so they can see in the dark, but they also use their ears. Did you know an owl has one ear higher than the other? Why do you suppose that is? I'll show you.

Let's do a little scientific research. First, you need to formulate a hypothesis: why do you think one ear is higher than the other? Everybody pick a partner and discuss this. Next, I need a control group. [*Choose five students.*] You five keep your eyes open, so you can watch me clapping. I would also like you to help me monitor this experiment by observing your classmates.

The rest of you are my guinea pigs. Close your eyes. Now I'm going to go around the room and clap, and you try to point to where the clapping sound is coming from. Use your ears. Don't cheat, do a little scientific test here. Don't open your eyes until I say so. Point at the clapping sound. [*Clap three times. Move to a different part of the room. Clap three times. Move your hands higher. Clap three times. Move your hands lower. Clap three times. Repeat this in various parts of the room.*] Okay, everyone can open your eyes, please.

How did you know where the sound was coming from? Do you have a hypothesis? [*Ask several students.*] How did you know where to point? You heard it over here because this ear heard it first. You heard it over there because that ear heard it first. Who can add to that answer, who has another idea? Yes, sound travels in waves and it take longer to get to the other ear.

This is how science works. Nobody has all of the answers; scientific discoveries are made in pieces. We all put our minds together and that's how we get the best answers.

So, why does an owl have one ear higher than the other? Let's use our research to answer the question. An owl has one ear higher than the other so it can tell where sounds are coming from—above or below.

Do you have a dog or cat? When you talk to it, does it turn its head? Cats and dogs do that to put one ear above the other, to create the same effect that an owl has naturally.

Rusty knew how to listen, and she knew how to call to other owls. Whoo-whoo, whoooo, whoo-whoo, whoo-whoo. Let me teach you the call of the great horned owl. Their long distance call is 1–2, 3, 4–5, 6–7, always an odd number. Listen to the rhythm, then you can try. Whoo-whoo, whoo, whoo-whooo, whoo-whooo. If you go out in the late evening or early morning, you can call an owl. They live in most cities.

But let's get back to your story.

As an owl, Rusty spent her nights catching snakes and mice, rabbits and skunks. (Yes, skunks! An owl does not have a sense of smell. If a skunk sprays it, the owl does not notice.) But Rusty grew old—just as you and I will one day. After a while her eyes were not so clear, and her ears didn't work so well. After a while she wasn't able to hunt very well, and her wings didn't work as well as they used to. She tried to catch a mouse, but it got away. She tried to catch a rabbit, but it was too fast.

One night late in October, Rusty the owl was hungry. Because she could not catch any food, she had no fuel. Because she had no fuel, she couldn't say warm. Rusty the owl shivered. She froze and died. She fell from the tree. She was buried in a blanket of reds and yellows as the autumn leaves fell upon her. The deep snow of winter covered her with a blanket of white. All winter long the owl lay there, until spring came and the snow melted. The leaves began to rot and so did the owl. It really began to stink.

Along came a fly. You might think that the owl smelled gross, but the fly smelled dinner! The fly flew down and

took a bite. Then the fly laid eggs. The eggs hatched. Have you ever seen maggots? They look like little grains of rice that wiggle. Those maggots ate some more of the decomposing owl.

Rusty did not became part of the fly; she did not become part of a maggot. Like most of the food and fiber you eat, Rusty was only passing through, if you know what I mean. This is an important part of science, part of the cycle of life. But let's move right along . . .

Then the rains came. The spring rains washed Rusty into the soil. Rusty became part of the mud—common, ordinary, everyday mud.

Just like you. You are made of mud. Do you eat food? Where does that food grow? It grows in the dirt, in mud. So, if you are what you eat and it grows in the mud, from what materials are you made? You're made from mud, but you're a lucky pile of mud. You can stand and walk. You can write poetry and play basketball. Most mud just lies there, like some people you may know.

But Rusty didn't lie there for long. Growing nearby, on the edge of the forest, on the edge of a field, was a prairie clover. The root of this plant sucked Rusty in. Slurp! Just like the oak tree did. Prairie clover plants need iron, too. So, Rusty traveled up the root of the prairie clover to its stem, and was deposited in the petal of the flower. The flower was reddish purple in part because of the iron and other minerals in the petal.

Then along came a butterfly. Many insects are attracted by the bright colors of flowers and the sweet scent of nectar. The butterfly landed there on the flower petal and unrolled its long proboscis, a long, hollow, straw-like tongue. The butterfly sucked up the flower's nectar. Rusty was part of the flower's petal—not the nectar—so when the butterfly flew away, Rusty was still in the flower.

Along came a buck, a male deer. This was a huge, 12-point buck. This deer was hungry. He saw that red clover and chomp, chomp, chomp, he ate it. Rusty became part of the buck.

Rusty loved being a buck. He loved running through the forest and doing all of the things deer do. He ran and jumped over logs. He used his antlers to fight with other male deer to tell them, "Hey, leave my does alone."

But later that year, when Rusty was coming down a trail, he passed a hunter with a bow. The hunter was hiding in the bushes along the edge of the trail. He waited beside the trail where he had seen Rusty go every day. As Rusty approached, the hunter began to quietly sing a sacred song that called the deer. What the song means is, "Native deer come join the feast, black-nosed deer come join the feast." Rusty heard that song.

When Rusty heard that song, he walked toward the hunter. The hunter readied an arrow, but before he fired he looked the deer in the eye and with a silent prayer asked, "Would you give your life so the people may live?' The deer was just a few yards off, but if he had turned to walk away, the hunter would have let him go. The hunter asked again in a silent prayer, "Would you give your life so the people may live?" Think about it. Would you give your life?

The deer turned as if to give the hunter a better shot, and a great shot he was. He shot his arrow right into the shoulder, right through the lung and into the heart. The deer jumped once and fell over dead.

Now, different tribes do this differently, but what the hunter may have done next is taken some of the corn he'd grown and sprinkled that corn on the mouth of the deer. An offering to the spirit of the deer would be made so the spirit of the deer would come back and there would be deer again next year. What he may have done next is taken a tool and cut the belly of that deer open. He would have taken out some the organs because they'll spoil the meat, but the liver and kidneys, these he saved for his grandmother and grandfather because they were delicacies. He may have eaten the heart while it was still warm. This way, part of the deer's heart would become part of his heart, and he would know the love and compassion of the deer.

Now different tribes do this differently, but what he may have done next is put that deer on his shoulder and carry it home. Because that deer gave its life, it was the hunter's duty to give it away. The hindquarter he may have given to his friends. The ribs he may have given to his sister. The front quarter he gave to the medicine woman, and Rusty became part of the old woman. But this was not any old woman. This old woman knew every plant in the forest.

Would you like to know that? Would you like to know every plant that grows around here? Which ones are poisonous, which ones are medicinal, for healing; which ones you can eat, like berries? Rusty knew how to make baskets, how to tan hides and make clothes. As a midwife, Rusty helped many babies come into this world and helped many elders cross over to that other world. Finally it was Rusty's time, and one morning the old woman did what you and I are going to do one day—she didn't wake up. Her people, the people of the river, burned down her house on top of her and buried her in a mound with her earthly possession. That is how the river people buried their dead. That is why you find small earthen mounds throughout the Midwest.

Let me ask you this: Would you like scientists to dig up your bones for research? Or, would you like a grave robber to dig up your grandfather's bones and sell his jewelry and other possessions? No? When you see these mounds, respect them as sacred sites, the resting places of elders.

So the cycle went on. Years later, a chipmunk buried a hickory nut in that mound. The hickory nut sprouted; it sent out a root that wrapped around the bone of the old woman. So rusty became a part of the hickory tree. Many years later, he became a hickory nut and was eaten by the great-great-grandmunk of the chipmunk who planted the tree. The chipmunk was eaten by a black snake, who was eaten by a red-tailed hawk. And so the cycle goes on. Rumor has it that Rusty is still out there. So on your way home, or as you go for a walk in the woods, keep your eyes open. You might run into him.

That was the story of Rusty. I told you this was the ballad of Rusty and Nancy. Nancy had a very different life. As a matter of fact, I would like to challenge you to a little higher level of thinking called comparative analysis. How is Nancy's life different from Rusty's? As you listen to this story try to figure that out. In many ways they are the same. How are they the same? How are they different?

To begin with, Nancy was not a solid like Rusty, like iron. Nancy was nitrogen, a gas. She was less tightly bound to other molecules—lighter, warmer, gaseous. Take a deep breath. Hold it. Let it out. Now take another deep breath. The air you just breathed in was inside the person next to you a few seconds ago.

Take a deep breath. What you just breathed in is mostly nitrogen. Air is 78 percent nitrogen and 21 percent oxygen. Can you imagine a graph of that?

It is oxygen we need to live. We get oxygen through our lungs when we breathe. When we exhale (who-o-o-sh), we give off carbon dioxide. Trees and plants are just the opposite. They take in carbon dioxide and give off oxygen. Plants don't have lungs of course; they don't breathe. Instead, they transpire gases through stoma (plural: stomata), tiny holes in their leaves.

Once more, take another deep breath. The oxygen you just breathed in might have been inside a tree a few weeks ago. Say, "Thank you, trees." Give thanks to the plants every time you breathe.

But let's get back to Nancy; Nancy was there when the universe was made. Well, to be more exact, the electrons and protons that nitrogen is made of were there. In the furnace of a star, these electrons and protons combined to form nitrogen atoms. Nancy was there. Nancy was there when Earth was made 4.5 billion years ago. Nitrogen was one of many gases, whirling and twirling in young Earth's atmosphere, an atmosphere that was too harsh to support life as we know it. Over billions of years, Earth's atmosphere changed. Nancy was there.

Nancy was part of the upper atmosphere. Nancy floated in the currents of air circling the globe. She was caught in a jet stream. A jet stream is a stream of air that swirls around the planet, around and around. Nancy flew over the supercontinent Pangaea. (Remember that?) She flew over North America and Asia as they moved apart. She flew around and around for a billion years, around and around, around and around.

But Earth was changing. I told you the continents were moving, right? North America moves—how much? One inch each year. Imagine this: Alaska was once down near the equator! Land that is now freezing cold was once a tropical swamp. Can you imagine that?

This was back before the days of flowering plants. There were plants, all right, but there weren't any flowers. Plants like ferns and mosses produce spores, not seeds, and they reproduce through those spores.

You've all seen little mosses and ferns growing in the forest? Well, in those days, hundreds of millions of years ago, the ferns were as big as trees. The moss was as big as a bush. Those were the days before dinosaurs, before reptiles, before human beings. In those days, insects controlled the world. There were dragonflies nearly as large as a red-tailed hawk! The world was theirs.

In that huge swamp the soil, the mud, the water was slightly acidic, like a lemon. The water was a weak acid, so many microscopic organisms couldn't live in it. The acid killed the tiny animals. There was no bacteria, there were no germs.

Now, if you took a handful of rich garden soil, you would find more than 1,000 organisms living in that one handful. But in that swamp, no organisms grew. Because it was acidic, no bacteria could live there.

Plants aren't microorganisms. Plants don't mind a little acid. So, they grew in that swamp. Everywhere you looked there were giant ferns and mosses.

Let's get back to Nancy. Every living thing needs nitrogen. Take a deep breath. Air is 78 percent nitrogen, but you can't absorb it through your lungs by breathing. Most plants can't absorb nitrogen from the air either. Like us, plants get their nitrogen from their food supply. Where do plants get their food? Well, one place a plant gets nutrients is the soil; they absorb nutrients through their roots. But if there is no nitrogen in the soil, how do plants get it?

Well, this huge fern growing in the swamp relied on some blue-green algae. The blue-green algae lived in a puddle on that fern. The algae lived in a little puddle that formed down where the leaf meets the stem. It lived there because rain water collects there, and the rain water is not acidic. Even in difficult situations, living things adapt and thrive! Blue-green algae can absorb nitrogen from the air. There was plenty of nitrogen in the air, so the blue-green algae had all the nitrogen it needed and more. With nitrogen to spare, the algae had something that the plant needed. But the plant also had something the algae needed: sugar. Remember, in "The Ballad of Rusty," we talked about how the factory in the leaves turn sunlight into food—sugar and starch? Well, a fern is all leaf—it makes plenty of sugar and starch. So there was the algae with

nitrogen to spare, and there was the fern with sugar to spare. What did they do? They found a way to exchange sugar for nitrogen. Can you say *symbiosis*? It means two things live together, and each helps the other survive. In this case, the fern gave up some sugar or starch, and the blue-green algae gave up some nitrogen. So, the fern traded a little of its sugar for algae's nitrogen.

Now let's get back to Nancy. Nancy moved down from the upper atmosphere. She cooled and settled closer to the surface of the Earth. Cool things sink, warm things rise. As Nancy settled, she was absorbed by that blue-green algae. *Zoop!* The algae passed her on to the fern. How did that happen? The fern could absorb moisture and the nutrients dissolved in the moisture through its leaf whorl, or base. *Slurp!* Nancy became part of that fern.

Eventually, that fern fell over and died. *Splat!* It fell into the acidic mud. But the fern didn't rot the way most things do. It takes living organisms—like bacteria, worms, and maggots you find in garden soil—to make things rot, and this acidic swamp had no bacteria. Because there was no bacteria to break it down the fern just slowly dissolved. Other ferns fell on top of it, *Splat!* and it was buried there in that mud. All together they dissolved and made a thick, oozy mud puddle.

Can you imagine a mud puddle the size of Chicago, Illinois? In some places that puddle was more than 100 feet thick! Slowly the continents drifted. *KABOOM!!!* Volcanoes erupted and lava poured over that mud. Now the mud puddle was encased in rock. The continent the swamp was on drifted to the north—drifted so far, it ended up in a very cold place. I told you where before. Do you remember? That's right! Alaska! The continent drifted north, farther and farther, and today the land that was once a tropical, fern-filled swamp, is in Alaska!

For a while the land was under water. Year after year dead sea creatures piled up on top. Layer after layer of sand and sea shells covered the lava that covered Nancy. How do I know? At the top of Denali, or Mount McKinley, in Alaska, the tallest mountain in North America, we find sea shells. Isn't that amazing?

So the lava and limestone and soil and rocks piled up. Nancy was crushed, compressed under 1,000 feet of rock. Can you imagine what would happen if someone took a rock more than 1,000 feet thick and set it on top of us right now? We'd be flat! That's what happened to Nancy, to all of that mud. It was compressed under lots of stone—sandstone, limestone. All of that rock squeezed the mud. All of that pressure changed the mud. Molecules, smashed together, began to combine in new and unusual ways. They formed oil. Over millions of years, the whole mud puddle turned into one huge, underground lake of oil.

A few million years later, give or take 100,000 years, there was a scientist, a geologist. A geologist is a person who studies the history of Earth through its rocks. This is pretty fascinating stuff! Imagine learning to read history in a rock, looking at a mountain or a canyon, and being able to see what happened there over millions of years. That's what geologists do. This geologist had a choice: he could work at an environmental education center and teach people about Earth and geology, or he could travel the world, looking for oil. This geologist explored Earth looking for oil until finally, he went on an expedition to Alaska.

The geologist sent a pipe down into the Earth and he struck it rich. They don't call it black gold for nothing! After resting for millions of years, Nancy was drawn out of the ground. *Slurp!* A huge pipeline was built. This pipeline is so big that 10 of us could walk through it side by side. This pipeline is so long, it goes all the way from northern Alaska, from Prudhoe Bay in the Arctic Circle, to southern Alaska, Prince William Sound, near Valdez.

All of that oil is sucked out of the ground and pumped into the pipeline. It flows through the pipeline to Prince William Sound, where it is loaded into ships. The ships sail straight out of the bay across the Pacific Ocean.

Sometimes the ships travel straight across the Pacific Ocean without any problems. But sometimes, they develop a leak or they hit something like a reef and break open. Then the oil spills out and floats in a thick layer on the ocean and you know what happens then: carnage and catastrophe, death and destruction.

But let's get back to Nancy. At Valdez, Nancy was loaded onto a ship that went straight to Hong Kong with no leaks or accidents. In Hong Kong, she was unloaded at a refinery, where she was heated up and mixed with other chemicals. Molecules and atoms were tied together in a new and unusual way. Nancy became a new, man-made material called plastic. We call it man-made because the molecules and atoms are tied together in a way that is never found in nature.

After the refinery Nancy was shipped to a factory, where she was made into a little red boat. That little red boat was put in a box that was wrapped with shrink-wrap (another kind of plastic). Hundreds of boxes holding little red boats were loaded onto a crate, and then the whole crate was wrapped in more shrink-wrap.

Then it was time for another ride. A big crane loaded the crate onto another ship. That crane burned diesel fuel, which is a petroleum product—which is made up of Nancy's friends.

The ship—which also burned diesel fuel, which is a petroleum product, which is made up of more of Nancy's friends—motored back across the Pacific Ocean to the United States. Finally Nancy came back to America! At the docks, another crane unloaded the crate of little red boats. That crane also burned diesel fuel—which is a petroleum product, which is made up of more of Nancy's friends. [*At some point the audience will naturally begin to chant this little chorus with you—more of Nancy's friends!*]

The truck carried Nancy from Seattle to Chicago. The truck burned diesel fuel—which is a petroleum product, more of Nancy's friends. In Chicago she was taken to a warehouse, where she was unloaded by a forklift. The forklift burned diesel fuel—which is a petroleum product, which is *more* of Nancy's friends.

Nancy was stored in the warehouse for nearly a year. That warehouse was heated with kerosene heaters. Kerosene is a petroleum product, which is made up of—right! *More* of Nancy's friends. Finally, Nancy was loaded onto another diesel truck, which drove to a store in the heart of a city. The truck burned diesel fuel—which is more of Nancy's friends. At the store, she was unloaded and stocked on the shelves.

The next day, a mother was shopping with her son. The little boy saw that red boat and he said, "Mommy, Mommy, please? I've been a good boy. Can I have that red plastic boat?"

And his mother said, "No. You have enough toys already."

But the boy took one of the boats and carried it with him anyway. When they went to pay for the things they were buying, the mother saw the boat and said, "No" and put it back on the shelf. The little boy was sad because he really wanted that boat.

Now, that mother was pretty smart. Her son's birthday was coming up, and she remembered that red plastic boat. So, one Saturday morning she took her little boy to grandma's house while she went shopping. Grandma was cool; she took the little boy to a park, where they had a great time and saw—guess who?—Rusty!

Mother went back to the store. She bought that red plastic boat and wrapped it in a beautiful wrapping paper with a gorgeous ribbon and bow. On her son's birthday she threw a big party. All of his friends came over and they sang (are you ready?), "Happy Birthday to you, happy birthday to you . . ." They played pin the tail on the donkey and all of those funny old games. Do you still play those games? All the kids had a great time, and then the little boy got to open his presents.

He got a baseball from one friend. "Hey, thanks!" Another friend bought him a bat. "Cool!" A third friend bought him a glove. "You must have talked about it, thanks!"

Next he opened the present he got from his mother. He took off the ribbon and bow. That ribbon was made out of polyester—which is a petroleum product, which is made from more of Nancy's friends. And the gift wrap was printed with ink—which is a petroleum product, which is more of Nancy's friends. And the shrink wrap was more plastic—which is a petroleum product, which is made from *more* of Nancy's friends.

The little boy had the ribbon and the wrapping and the shrink-wrap and the box. That's a big pile of garbage for a little red plastic boat.

As soon as the party was over, the little boy and his friends went to a creek behind their house. They were sailing his little boat down that creek when the little boy had an idea. He said, "Hey, let's pretend it's a war." One of his friends said, "I'll be the bad guy." As the little boat went down the creek, they tried throwing rocks at it. One kid picked up

The Ballad of Rusty and Nancy

a big rock and—*Splat!* A direct hit! That little red plastic boat, which the boy had only had for about one hour, was in pieces floating down the creek.

Nancy—as the plastic molecule— broke off of a tiny frag- ment. Because she was part of a chemical that nature had never seen before, she did not blend in too well. When a catfish came along it was curious. 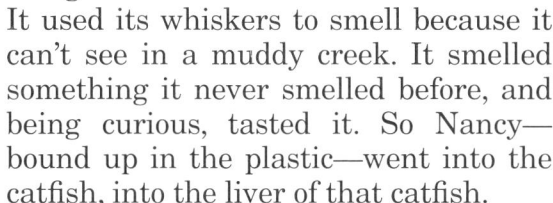 It used its whiskers to smell because it can't see in a muddy creek. It smelled something it never smelled before, and being curious, tasted it. So Nancy— bound up in the plastic—went into the catfish, into the liver of that catfish.

Because Nancy had been changed into a chemical that was new to the nat- ural world, the catfish's liver couldn't handle it. The poisons and the toxins built up, and one day the catfish died.

It turned belly up, and it floated down along the edge of the creek, where I think it might still be stinking.

And what happened to Nancy then? Nancy had been changed into an unnatural product, and she could never go back again. So, eventually the cat- fish rotted away, but Nancy remained im- prisoned in the plas- tic molecule forever.

You should know. . . . We have a choice. Every time we buy something or make something, we make a choice: What do we want more of, Rusty or Nancy? Do we want to nourish the web of life, or do we want to disrupt the bal- ance, change the natural order? What do we want Earth to be like in 100,000 years? It's up to you and me and all of us. What do we want more of, Rusty or Nancy?

Discussion Topics: Compare and Contrast

Immediately following this trilogy of stories, discuss the differences between and similarities of Rusty and Nancy. On the chalkboard or white board draw two overlapping circles (a Venn diagram). In one circle list the characteristics unique to Rusty. In the other circle list the characteristics unique to Nancy. In the overlapping section list the characteristics that Rusty and Nancy share.

These stories compare natural cycles to the linear use, abuse, and disposal of man-made products. Discuss how the facts in the stories support this theme. Using facts from the stories, discuss ways that people are part of the circles of nature and ways that people have a negative impact on these circles. Brainstorm ways in which we could have a more positive impact and reduce our negative impact on the environment.

The Circular Storytelling Game

❖**Grade Levels**: K–12   **Time estimate**: 60 minutes

❖**Science skills**: Classification; Communication

❖**Objectives**: Students will demonstrate storytelling skills in the inventive use of scientific nonfiction. They will elaborate on concepts like the food web and predator-prey relationships through storytelling.

National Standards

Science Standards

NAS 1 Science as Inquiry: Abilities necessary to do scientific inquiry; Understandings about scientific inquiry.

NAS 3 Life Science: Structure and function in living systems; Reproduction and heredity; Regulation and behavior; Populations and ecosystems; Diversity and adaptations of organisms.

NAS 4 Earth and Space Science: Structure of the earth system; Earth's history; Earth in the solar system

Language Arts Standards

NCTE 4 Students adjust their use of spoken, written, and visual language (e.g., conventions, style, vocabulary) to communicate effectively with a variety of audiences and for different purposes.

NCTE 5 Students employ a wide range of strategies as they write and use different writing process elements appropriately to communicate with different audiences for a variety of purposes.

NCTE 6 Students apply knowledge of language structure, language conventions (e.g., spelling and punctuation), media techniques, figurative language, and genre to create, critique, and discuss print and nonprint texts.

NCTE 11 Students participate as knowledgeable, reflective, creative, and critical members of a variety of literacy communities.

NCTE 12 Students use spoken, written, and visual language to accomplish their own purposes (e.g., for learning, enjoyment, persuasion, and the exchange of information).

❖**Materials**: Pen or pencil and paper; chalk

Instructional Procedures

Introduction: Many of us have played a simple storytelling game in which one person begins a story, then suddenly stops and hands the story off to the next person. The second person picks up wherever the first person left off, then suddenly stops and hands the story off to the next person. The story is passed from person to person until everyone has had a chance to add something to it.

Giving this old game a new twist, you can deepen students' understanding of many cycles of nature, such as the mineral cycle, the gaseous cycle, the food web, and the water cycle.

Activity: Use "The Ballad of Rusty and Nancy" as a model. Start to tell a story about the cycles of nature, then suddenly stop. Working in small cooperative groups of four students, they continue the story. There are two ways to continue the exercise: storytelling and story writing. I recommend using both activities in that order. Remember, oral language development is an important step in reading and writing. When students have a chance to tell, then listen to, other versions of a story, it enriches their understanding of the concepts and process of creating stories. When they read and write stories in this circular game, they see several models and indirectly teach each other to write or create stories. I have seen immediate and dramatic improvement in the clarity, detail, flow, and complexity of students' writing as a result of this simple two-part exercise.

Storytelling

You can pass the story orally, with each student giving some specific details about one life form. For example, you could start a story about Rusty, who became part of an alfalfa plant and was eaten by a rabbit. Then give details about Rusty as a rabbit. For example, as a rabbit Rusty loved to hop around, wiggle his nose, and eat wild greens. He was fast, he had big hind feet, and in each season his coat changed colors. One day he crawled out of his hole in the ground, and as he was crossing the snow-covered meadow, *snatch!* What happens next? Hand the story off.

The next storyteller might tell how Rusty was eaten by a fox, then add details about Rusty the fox. For example: Rusty loved being a fox. She had a long, slender nose that could sniff out her enemies and prey. She had a big bushy tail that helped her to keep her balance as she ran. She also used her tail to keep warm on a cold winter night. Each student adds detail about Rusty's current life form. The more detail you model when you start the story, the more detail you will elicit from your students.

Story Writing

Next, you can do the story circle as a writing assignment. Assign students to small groups of about four students each. Ask each student to write the beginning of a story on a sheet of paper. (All four students do this, each one writing a different beginning.) After an allotted time, the students stop writing and pass their story to the student on their right. Each student takes a moment to read what his or her teammate has written and then adds to the story. This process repeats until all the students in each group have had a chance to contribute to each story. Allow 3–10 minutes for each writing interval, depending on your students' grade level, writing ability, and attention spans.

Combining Writing and Telling

Telling and writing can be combined in circular stories. Combining telling and writing has many benefits. Allowing students to listen to each other and to tell their stories informs the writing process and encourages fluid, cohesive writing. In addition, telling their stories or sharing what they have written with the class builds students' confidence and prepares them for telling an entire story by themselves. It is also a chance for them to learn from their peers through indirect, nonverbal feedback about what works and doesn't work in performance.

Individual Writing Exercise

After students have participated in circular storytelling and/or writing to stimulate their thinking, have each student write a version of "The Ballad of Rusty" or "The Story of Nancy." Have each student pick an element (iron or nitrogen) and then map out three or four incarnations or life forms in which the element might appear. Encourage students to go into great detail about each life form. Give students time to go to the library, go online and surf the Web to research details. As a guideline, suggest students describe one complete cycle. For example, a student could follow Rusty from mud to a plant, to a herbivore, to an omnivore, to a carnivore, and back to mud. However, remind them it is better to provide a great many details on one life form rather than race through three or four. Be sure to allow plenty of time for prewriting, freewriting, rewriting, and editing. The final drafts of their stories could be illustrated and collated into a classroom anthology.

Depending on your curriculum, these stories can focus on a wide range of science topics, including Rusty in a Rain Forest or Rusty the Single-Celled Organism. Challenge students to write creatively about things they have recently studied in science class.

Assessment: These stories can be graded based on the science concepts you are emphasizing as well as grammar, spelling, and punctuation. I often give students two grades for one assignment, a two-for-one bargain!

Follow-Up Activities: If everyone begins and ends with mud it becomes very easy to tie the individual students' stories together to make a novel about Rusty or Nancy. With simple word processing and desktop publishing this can be grafted together with illustrations to make a beautiful book. Or they could be saved into a blog format so students can contribute stories as the year goes on, and as you explore different science content students could be encouraged to add to the story with a different emphasis. For example, early in the year you could focus on Rusty as a geological story, later in the year it could be Rusty the Red Blood Cell to explore internal anatomy.

Exploring the Effects of Our Behavior: Monitoring Waste

❖**Grade Levels**: K–12

Time estimate: 30 minutes for the initial exercise, which could lead to a lifetime of best practices!

❖**Science Skills**: Classification; Metric Measurement; Communication

❖**Objectives**: Students will demonstrate an understanding of metric measurement and their impact on the environment by keeping track of individual and classroom waste.

National Standards
Science Standards

NAS 1 Science as Inquiry: Abilities necessary to do scientific inquiry; Understandings about scientific inquiry.

NAS 6 Science in Personal and Social Perspectives: Personal health; Populations, resources, and environments; Natural hazards; Risks and benefits; Science and technology in society

Language Arts Standards

NCTE 4 Students adjust their use of spoken, written, and visual language (e.g., conventions, style, vocabulary) to communicate effectively with a variety of audiences and for different purposes.

NCTE 5 Students employ a wide range of strategies as they write and use different writing process elements appropriately to communicate with different audiences for a variety of purposes.

❖**Materials**: Trash; plastic gloves; kitchen or bathroom scale; pen or pencil and paper

Instructional Procedures

Introduction: As "The Ballad of Nancy" makes clear, every day each of us makes decisions that adversely affect the environment. It can be enlightening to catalog the ways each of us affects the environment.

Activities: As a class, sift through a bag of trash to find recyclable materials, organic waste for composting, or unnecessary packaging. What classifications do the students come up with? Wearing plastic gloves, students sort the trash into piles, then use a scale to weigh the amount of each type of waste. They chart or graph the results. (To be sanitary and safe, you may want to create and bring in a sample bag of garbage.)

As a homework assignment, ask each student to sift through a bag of family garbage and produce a graph showing the amounts of materials that could be recycled, compostable organic waste, unnecessary packaging, and other types of trash they find. The students compare their graphs. Emphasize safety and sanitary practices, and encourage parental supervision. After students have a chance to inspect the trash, talk about ways to reduce waste.

Discuss bulk purchases, packaging that is recycled or that can be recycled, composting, using cloth bags instead of paper or plastic at the grocery store, and other options for reducing waste.

Expand your trash inspections beyond the classroom. Monitor food waste and excess packaging in the lunch room. After students document the waste in the lunch room, talk about ways to reduce it. Encourage students to bring their lunch in reusable containers and to bring only what they will actually eat. Some schools have eliminated up to 70 percent of their waste!

Waste at Home and on the Road

Junk mail. Junk mail is a huge waste of natural resources. Junk mail wastes not only trees (to make paper) but also petroleum (to make ink, plastic wrap, coatings on paper, and the fuel needed to ship it across the country). To monitor this waste, families can collect all of the junk mail they receive in a month, then weigh it and multiply that weight by 12. This will yield the number of kilograms of junk mail the family can expect to receive in one year. To reduce this waste, the family can write to the companies involved and ask to be taken off the mailing lists. (Send the request to www.dmachoice.org/ or Direct Marketing Association, 1120 Avenue of the Americas, New York, NY 10036–6700.)

Road hog. Have students monitor the number of miles they travel by automobile in an average week. Then discuss ways to reduce their auto miles by walking, bicycling, riding mass transit, and making more efficient use of each trip, for example, by combining errands.

Resources. Have students list all of the petroleum products they use every day. Then discuss renewable versus nonrenewable resources. Have students talk about ways to make more efficient use of nonrenewable resources and ways to use renewable resources in place of nonrenewables. In addition, students might come up with ways to reduce their use of resources altogether, for example, they might turn off the water while they brush their teeth or carry their lunch in a cloth bag instead of a paper sack.

There are a growing number of websites that help folks monitor their carbon footprint, estimate water consumption, and in general be more conscious about their personal impact on the earth and how to make a difference. It really does begin at home. *And* these kinds of activities are empowering: to know we can make a difference one household, one school , one community at a time. Here are a few of my favorite sites:

This simple calculator does require some information up front. I found it easier to skim it so I know what data to collect and then come back to fill in the blanks: http://www.carbonfootprint.com/calculator.aspx.

The Nature Conservancy Carbon Footprint Calculator is easier to use, asks more general than specific questions, and gives positive, easy-to-accomplish suggestions for reducing your impact: http://www.nature.org/initiatives/climatechange/calculator.

Assessment: This could be a simple math work sheet where students turn in word problems that outline the amount of waste they measured and bonus points offered for the amount of waste they reduced over time.

Follow-Up Activities: Fired up about the waste they are reducing, students could be challenged to reduce school waste, measure the amount of savings to the school budget, and write persuasive essays to the principal or school board.

Social Action and Citizenship

❖**Grade Levels**: K–12　　　　　　　**Time estimate**: Weeks or Years!

❖**Science Skills**: Observation; Communication; Identify Variables; Design Investigations

❖**Objectives**: Students will exemplify the insight, commitment, and problem-solving necessary to be an active member of our democracy!

National Standards

Science Standards

NAS 1 Science as Inquiry: Abilities necessary to do scientific inquiry; Understandings about scientific inquiry.

NAS 3 Life Science: Structure and function in living systems; Reproduction and heredity; Regulation and behavior; Populations and ecosystems; Diversity and adaptations of organisms.

NAS 6 Science in Personal and Social Perspectives: Personal health; Populations, resources, and environments; Natural hazards; Risks and benefits; Science and technology in society

Language Arts Standards

NCTE 1 Students read a wide range of print and nonprint texts to build an understanding of texts, of themselves, and of the cultures of the United States and the world; to acquire new information; to respond to the needs and demands of society and the workplace; and for personal fulfillment. Among these texts are fiction and nonfiction, classic and contemporary works.

NCTE 4 Students adjust their use of spoken, written, and visual language (e.g., conventions, style, vocabulary) to communicate effectively with a variety of audiences and for different purposes.

NCTE 5 Students employ a wide range of strategies as they write and use different writing process elements appropriately to communicate with different audiences for a variety of purposes.

NCTE 7 Students conduct research on issues and interests by generating ideas and questions, and by posing problems. They gather, evaluate, and synthesize data from a variety of sources (e.g., print and nonprint texts, artifacts, people) to communicate their discoveries in ways that suit their purpose and audience.

❖**Materials**: *The Lorax*, by Dr. Seuss; newspapers; pen or pencil and paper and whatever else students determine is needed: shovels, banners, petitions, e-mail campaigns, etc.

Instructional Procedures

Introduction: From the Boston Tea Party to the suffrage movement, from the end of slavery to the modern struggle for civil rights, one of the most important parts of our democracy is this: Americans have stood up for what they believe in and did whatever was necessary to improve the lives of their fellow citizens. One of the strengths of the environmental movement is that it gives average folks an opportunity to be a part of this most American legacy!

Activity: Read aloud *The Lorax* by Dr. Seuss and then, as a class, discuss local environmental problems. Encourage your class to be like the Lorax, a voice for the earth, and before the last Truffula Tree is gone, do something to make the earth a healthier place. Through Socratic questions challenge students to first list possible problems and then explore possible solutions. As a class, vote on one local environmental problem. Ask students to bring in information, surf the Web, read recent newspaper articles online, study the root causes, and discuss what others have done to solve the problem. Visit the site of the problem and interview officials and neighbors. Stage a debate with students arguing pro and con. Explore ways in which students can be part of the solution. Arrange to have students help with the cleanup or remediation at the site. Have the class write letters to Congress or local corporate executives encouraging them to act responsibly. Study the workings of Congress and how a bill becomes a law so that students understand how their letters affect legislation.

Earth Day Every Day, although a great idea, has become a cliché. Still, it is important to be mindful of the ways we affect the environment, both positively and negatively, with our every action. More important, we all need to look at the implications of our behavior and make decisions that will ensure a healthy environment for the future. Much of the action taken to protect the environment is reactive, a response to problems humans have created. This level of work is necessary and urgent, but it is also important to be proactive, bold, and visionary.

Brainstorm a list of projects you could undertake to improve your local habitat. Following are a few ideas to start the conversation:

- Plant a tree at school and at home.
- Plan, plant, and maintain a butterfly garden, hummingbird garden, wildflower garden, or schoolyard or backyard habitat.
- Develop a nature trail at school or at a local park.
- Help convert an abandoned railroad grade to a hiking/biking trail.
- Plant fruit and nut trees for people and for wildlife.
- Arrange a class field trip to a local nature preserve.

Chose one project and then go outside to celebrate your relationship to the wild world.

Assessment: This is the kind of project that offers multiple levels of assessment. If they are writing letters to congress or editorials for the local paper, their persuasive writing could be evaluated. If they are planning and planting a garden or building a nature trail, students could get graded on their layout and design using geometry, the math of their budget, and the economics of labor and community enrichment. (Property values go up when a neighborhood gets a new rail-to-trail hiking/biking trail!) As the project develops, the teacher can set standards and inform students as to expectations and how they will be assessed.

Follow-Up Activities: Spend a lifetime enjoying the fruits of your labors!

Bibliography for Further Research

For an excellent introduction to the science behind the big bang theory, please visit: http://www.talkorigins.org/faqs/astronomy/bigbang.html or you can type "evidence for the big bang theory" into any search engine.

For a hands-on research project that explores geologic evidence for continental drift, visit the Discovery Channels page of lesson plans: http://school.discoveryeducation.com/lessonplans/programs/continentaldrift/

A well-designed lesson that uses map skills to explore the supercontinent can be found here at Pangaea Puzzle: http://www.aktsunami.com/lessons/k-4/unit4/atep_k4_PangaeaPuzzle.pdf

For more information about owl hearing, here are two excellent Web pages: http://www.owlpages.com/articles.php?section=Owl+Physiology&title=Hearing and http://videos.howstuffworks.com/discovery/30497-the-ultimate-guide-birds-of-prey-barn-owl-hearing-video.htm

Baylor, Byrd. *If You Are A Hunter of Fossils*. Aladdin Books, 1980. 13: 978–0689707735. Like all of Baylor's highly recommended books, this is a gentle guide into inquiry and listening to the stories of the land, encouraging readers to think about the stories held in stone.

Deedy, Carmon Agra. *Agatha's Feather Bed*. Peachtree, 1991. 13: 978–1561450961. This hilarious look at the connections between what we use and where it comes from achieves that rare balancing act of being delightful for kids and inspiring for adults.

Guiberson, Brenda, illus. by Gennady Spirin. *Life in the Boreal Forest*. Henry, Holt, 2009. 13: 978–0-8050–7718–6. This book is a work of fine art and deep ecology. Rarely do words and illustrations work so well together to create an immersive experience in drawing readers into an intimate experience with the wild world.

Lorbiecki, Marybeth, illus. by Nancy Meyers. *Planet Patrol: A Kid's Action Guide to Earth Care*. Two-Can, 2005. 13: 978–1587285141. One of the more attractive and practical guides to making a difference with lots of simple and step-by-step complex ideas for improving our local environment. I also learned some cool facts, like this one: planting a $5 tree today will yield a $191,000 return on your investment!

Morgan, Jennifer, illus. by Dana Lynne Andersen. *Born with A Bang*. Dawn Publications, 2002. 13: 978–1584690320. *From Lava to Life*. Dawn Publications, 2003. 13: 978–1584690429. *Mammals Who Morph*. Dawn Publications, 2006. 13: 978–1584690856. Dawn Publications. This series of three picture books, collectively known as *The Universe Tells Our Story*, does a delicate job of both encouraging a sense of awe and being scientifically accurate.

Walter the Water Molecule, by Brian "Fox" Ellis and Garth Gilchrist

I bring fresh showers for the thirsting flowers, / From the seas and the streams.

—Percy Bysshe Shelley, from *The Cloud*

Comments to the Teacher

THIS IS THE STORY OF THE WATER CYCLE. Similar to "The Ballad of Rusty and Nancy" in its cyclical nature, it covers the evaporation, condensation, and precipitation of our most precious natural resource.

This is another story that is easy to adapt to your local climate, geology, geography, and environmental conditions. Start in the lake or ocean as the source of rain. Trace Walter through your local weather patterns into your local rivers, lakes, and water supplies. Change the details to explore your local environment.

It is also very easy to tell a 5-minute or 45-minute version of this story, stretching or shortening to fit your time frame and adapting to your audience. I have told this story to preschool students and to an international wetlands conservation conference with a room full of PhD aquatic ecologists, simply by changing the vocabulary and amount of science details.

And now the Story . . .

Walter the Water Molecule

Once I was drinking a tall glass of water. Glunk! Glunk! Glunk! When I was finished, there was a tiny drop of water at the bottom of the glass. The sunlight was shining through the glass and making a little rainbow there on that drop of water. Looking at that drop of water, I thought about where it had been and where it was going and all the adventures that drop of water had seen.

You know that the water on Earth today has been here for billions of years, right? The same water that is in you today was inside a dinosaur 200 million years ago. As I looked at the drop of water, I began to think about its story, the story of the water cycle, the story of Walter the Water Molecule.

Walter's story begins at the bottom of the Pacific Ocean, deep, deep down, where the sun doesn't shine. Sunlight cannot make it so far down—more than a mile deep, down at the bottom of the ocean.

Way down there at the bottom of the ocean, Walter, a tiny drop of water, was squeezed in with a lot of other drops, billions of drops of water. There was water up above, and water down below. There was water in front, and water behind, water on the left and water on the right. Walter was squeezed in on every side. Can you imagine that? Squished! It was pretty boring most of the time. And dark! The sun's light cannot penetrate a mile into the ocean. There wasn't much to do. There wasn't much life down there. Oh, once in a while a glow-in-the-dark fish would come by, and that was exciting, but most of the time there wasn't much happening.

All that was about to change! Near the surface of the ocean was a giant sperm whale—one of the largest mammals on Earth. Whales breathe air like you and me. Take a deep breath. Hold it, hold it. *How long can you hold it?* You can hold your breath for a minute if you are lucky. A whale can hold its breath for nearly an hour. This gives it plenty of time to get down near the bottom of the ocean. This whale took a great big deep breath and dove down, down, d-o-o-w-w-n-n.

In the darkness of the ocean the whale could not see where it was going. How did it find its way? That's right, it *sang*. The whale sang: *wooohnng*! [Commercial recordings of whale songs are available at Ocean Mammal Institute: http://www.ocean-mammalinst.org/songs.html.] Have you ever heard a whale's song? Try singing that with me: *wooohnng*! The whale uses its voice like sonar or radar. It's called echolocation. The whale listens to the echo of its voice, and when that echo comes back, the whale can tell from the sound whether its voice is echoing off a ship (the whale needs to dive deep to avoid the harpoon), or a coral reef (the whale needs to avoid it because it will scrape its side), or a school of krill or small fish (dinner?). Most whales eat the tiny shrimp called krill, but a sperm whale will eat anything it can catch; giant squid is one of its favorites. This whale was hungry.

Looking for food, the whale dove deeper, deeper, down, down. It sang its song and listened. It heard the echo of something long and thin with a lot of legs. The object had more than eight legs, so it wasn't an octopus. It was a giant squid! The squid was swimming, trying to get away, and the whale was swimming, trying to catch the squid.

Next thing you know the squid wrapped its legs around the whale, and the two of them rolled and wrestled. As they rolled and wrestled they went farther and farther down, down to the bottom of the sea, where Walter floated, squished.

Chomp, chomp! The whale and giant squid fought. Chomp, chomp! The whale ate the squid. Chomp, Chomp! When the whale ate the squid it swallowed some water, too.

Walter was caught in the whale's mouth!

By now, the whale had been underwater for quite some time, so it had to race back to the surface to get a breath of air.

Walter was inside the whale! His adventure was finally beginning.

The whale reached the surface. You know that little blowhole in the top of the whale's head? Whoosh! Whales use that hole to breathe. You breathe through holes in your head, too: they are called your nose and mouth.

When whales breathe out, water comes out. So, when this whale breathed out, Walter rushed into the air, into the sunlight. Then *splash*! Walter landed on the surface of the ocean.

Walter rode a wave up, up, and down the other side. It was better than a roller coaster. U-u-u-p and d-o-w-w-w-n the other side. U-u-u-p and d-o-w-w-w-n. When Walter got to the top of one wave, he found drops of water jumping into the air. When he went u-u-u-p to the top of the next wave, he jumped into the air, too.

Walter had evaporated! Can you say *evaporation*? When water becomes airborne, it becomes water *vapor*, and that's e*vapor*ation!

The wind swept Walter up. It carried him higher and higher. Soon he was high above the Pacific Ocean. The wind carried Walter higher and higher. A few moments ago, Walter was a mile under the ocean; now he was a mile above the surface. Before long Walter floated up into a jet stream. Jet streams are rivers of air that circle the Earth. The jet stream carried Walter over the Pacific Ocean and over North America and across the Atlantic Ocean. And over Europe and the former Soviet Union. Across Asia and back over the Pacific Ocean. In a couple of days, Walter went around the entire planet.

Walter floated along in the jet stream for quite some time, going around and around the Earth. When he floated over Canada there was a high-pressure system that forced him down near the ground. This cold front pushed over Canada, so Walter sank. Cold air sinks, warm air rises. Not only did he sink down near the surface of the Earth, but he moved south toward the equator. As this cold front blew over the United States, a warm front moved up from the Gulf of Mexico. When the warm front and the cold front came together, they did a little dance called the thundercloud!

All of a sudden-*zuup, zuup!* — Walter was not alone. Tiny drops of water stuck to his front—*zuup!* —and his back—*zuup!*—and his left—*zuup!*—and his right—*zuup!* Before long, a cloud appeared there in the sky. Out of a blue sky, the moisture came together and a cloud was made. Can you say *condensation*? When the droplets come together, they condense. Walter was part of that cloud. A cloud is like a lake floating through the air.

Well, this cold front and warm front made more than just a cloud; they made a thunderhead, a huge, boiling cumulonimbus cloud filled with big drops of water. Soon Walter was too heavy to float. He passed the dew point, and he fell to the ground. Can you say *precipitation*? Walter was a raindrop! Whoa, whoa, whaaaaaa, *splat!* Walter fell to the ground with the other raindrops, left and right. There on the ground, the raindrops ran together. Walter ran into a little trickle, and this little trickle joined another trickle and formed a little creek. The little creek joined another little creek and made a little stream. This little stream, rolling over whitewater rapids and into deep pools, joined another stream and became a river.

Growing near this river was a huge sycamore tree. The tree had a root that reached down into the river. The root was like a straw. Slurp! The root sucked Walter into that tree. While he was down at the root, Walter picked up a little something: calcium. Then Walter went into the root, up through the trunk of that tree. Walter went up the trunk and into a branch of that tree. Walter went out on a limb and into a leaf on that tree. When he reached the leaf, the cells in the leaf traded some sugar for that calcium. Would you make that trade? If you had some calcium, would you give it to somebody for some tree sugar?

Walter carried the sugar back down the trunk of that tree. While he was carrying the sugar, it turned into starch! Walter carried the starch down to the root. When he got there, one of the cells traded Walter some iron for the starch. (Rusty?)

Walter was part of the transportation system of the tree. He was like a truck driver, carrying materials up and down the highway of the tree's trunk, driving through the xylem and the phloem cells, where moisture moves up and down from the root to the leaf and the leaf to the root.

One day, Walter came to another leaf and noticed little drops of water lining up next to a little hole. That hole is called a stoma. Basically, it's a hole in a leaf where a tree breathes, called *stomata* if there are more than one, and there are always more than one.

Walter joined the other drops of water lined up next to the stoma. The drop at the front of the line jumped out of the hole, and the line moved forward. Then the next drop jumped out of the hole, and the line moved forward again. *Aaah!* Out into the air Walter went. Can you say *evapo-transpiration*? It is a fancy name for tree sweat!

Yes, Walter had evaporated again! This time, though, instead of floating up into a cloud, he condensed onto a blade of grass. Sometimes, if you go out in the early morning after a cool night, you will see the grass is wet and maybe your bicycle is wet—but it hasn't rained. That moisture is called *dew* or *condensation*. Walter had become a drop of dew.

When the sun rose, most of the dew evaporated, but not Walter. He sank down into the ground, and this time he ran deep. The rains came and washed him down, down a worm tunnel, deep into the ground. This time, he became part of an underground lake, an aquifer. Hold on! There are lakes and rivers running underneath you right now! Hold on! Walter went into an underground river, the aquifer.

In this aquifer was a pipe. The pipe was part of a well. The pipe was a man-made straw. When somebody at the farmhouse on the top of the aquifer turned on the faucet, Walter flowed up through the pipe, through the well, through the faucet, and into somebody's glass.

That somebody was a little girl who was getting ready to go to school. She drank a tall glass of water. You're supposed to drink lots of water every day. Her mom said so. Gulp! Gulp! Gulp! Walter went into that little girl. She kissed her mom good-bye. She got on the bus and rode the bus to school.

Walter was going to school. All the kids laughed and played on the bus. When they got to school, Walter went through the brain of that little girl. She was smart! The water in her brain helped her with her math. From her brain Walter went down to her heart, which is the pump that moves the blood around. From her heart, he went to her lungs to pick up oxygen. From her lungs, he went to her heart, and then he went to her leg, where he helped the little girl walk to lunch.

From her leg he went into her heart, from her heart he went to her lungs, and from her lungs he went to her heart, and then he went to her stomach. Today was a good lunch: pizza, the little girl's favorite. The water in her stomach helped her to digest her lunch. Walter carried nutrients back to her heart, to her lungs, and to her heart, and then to different parts of her body. That is the way blood flows in your body: from your heart, to your lungs, to your heart, and then to parts of your body. Say this little rap with me: heart to the lungs, lungs to the heart, heart to the

body . . . body to the heart, heart to the lungs, lungs to the heart, heart to the body . . . body to the heart . . . faster and faster . . .

After lunch the little girl went outside for recess. She and her friends played her favorite game, kickball! When it came her turn to kick, she ran up and cracked that ball. *Whoosh!* It flew over second base. The little girl ran to first (huff, huff), she ran to second (huff, huff), she ran to third (huff, huff), and she ran all the way home (huff, huff). She slid into home base. Wow! A home run! Yea! Everyone clapped.

Huff! Huff! The little girl ran so hard, she worked up a sweat. When you sweat, you leak. You have tiny holes in your skin called pores. You can't see them because they're so small, but if you had a powerful magnifying glass, you could see lots of little holes in your skin. You are leaking more or less all the time. You are leaking right now. Every time you sweat you are leaking a lot!

Walter came out of the little girl as sweat, and then he evaporated again. This time he went into a cloud. The cloud blew across the United States and the cloud blew over the Atlantic Ocean, picking up moisture along the way. Eventually the cloud was thick with water. When the cold rain started to fall, *whoa, whoa, whaaaaaaa-splash!* Walter landed in the ocean. He floated down, down, down. A fish swam past, and Walter went into the fish. Walter went into the belly of the fish, into the blood of the fish, into the bladder of the fish. . . . What happened next? I will leave this to your imagination. Let me say that when the fish swam away, Walter was left behind.

Walter sank down, down, down, deeper and deeper. Before long he was at the bottom of the ocean. This time he was in the Atlantic Ocean, not the Pacific Ocean. But it was more or less the same. There was water in front and water behind; water on the left and water on the right; water up above and water down below. It was dark; the sunlight could not make it down so far. It was deep and dark, and there wasn't much to do . . . except to wait for the next adventure.

Follow-Up Ideas for "Walter the Water Molecule"

Discussion Topic

Immediately after telling the story, encourage students to think about where the water inside of them might have been. And where will it go when they are done with it? Ask them to turn to a partner and discuss this idea of the never-ending cycle of water and to take turns making up a story about Walter!

There are a few other versions of this story available, though Walter was first in print!

McKinney, Barbara Shaw, illus. Michael S. Maydak. *A Drop Around the World*. Dawn Publications, 2000. 1-883220-71-8. The best part of this book is the scientific back matter with four pages of hydrology found on the last few pages. The illustrations are great. The A-A, B-B rhyme scheme gets old, but the details of the story are rich as this drop of water literally travels around the world, in and out of plants and animals, and through the hydrological cycle. There is also a fine set of lesson plans available in a Teacher's Guide available from the publisher.

Morrison, Gordon. *A Drop of Water*. Houghton Mifflin, 2006. 978-0-618-58557-1. This simplified version of Walter, though it lacks scientific depth, has its strength in the easy sequential pattern and the astoundingly beautiful illustrations. It could be used as a graphic representation of the tale with added narration drawn from the pictures.

Correlating Maps and Diagrams

❖**Grade Levels**: 1–12   **Time estimate**: 45–60 minutes

❖**Science Skills**: Observation; Communication; Prediction; Inference

❖**Objectives**: Students will demonstrate an understanding of the water cycle though mapping the ways in which water moves through the earth, their watershed, and their bodies.

Students will exercise mapping skills and draw diagrams of the water cycle.

National Standards

Science Standards

NAS 1 Science as Inquiry: Abilities necessary to do scientific inquiry; Understandings about scientific inquiry.

NAS 2 Physical Science: Properties and changes of properties in matter; Motions and forces; Transfer of energy.

NAS 3 Life Science: Structure and function in living systems; Reproduction and heredity; Regulation and behavior; Populations and ecosystems; Diversity and adaptations of organisms.

NAS 6 Science in Personal and Social Perspectives: Personal health; Populations, resources, and environments; Natural hazards; Risks and benefits; Science and technology in society.

Language Arts Standards

NCTE 1 Students read a wide range of print and nonprint texts to build an understanding of texts, of themselves, and of the cultures of the United States and the world; to acquire new information; to respond to the needs and demands of society and the workplace; and for personal fulfillment. Among these texts are fiction and nonfiction, classic and contemporary works.

NCTE 3 Students apply a wide range of strategies to comprehend, interpret, evaluate, and appreciate texts. They draw on their prior experience, their interactions with other readers and writers, their knowledge of word meaning and of other texts, their word identification strategies, and their understanding of textual features (e.g., sound-letter correspondence, sentence structure, context, graphics).

NCTE 4 Students adjust their use of spoken, written, and visual language (e.g., conventions, style, vocabulary) to communicate effectively with a variety of audiences and for different purposes.

NCTE 5 Students employ a wide range of strategies as they write and use different writing process elements appropriately to communicate with different audiences for a variety of purposes.

NCTE 7 Students conduct research on issues and interests by generating ideas and questions, and by posing problems. They gather, evaluate, and synthesize data from a variety of sources (e.g., print and nonprint texts, artifacts, people) to communicate their discoveries in ways that suit their purpose and audience.

NCTE 8 Students use a variety of technological and information resources (e.g., libraries, databases, computer networks, video) to gather and synthesize information and to create and communicate knowledge.

❖**Materials**: The Water Cycle worksheet (page XX; photocopies of each of the following: an outline map of the world, an outline map of the United States, an outline map of your city, an outline of the human body; online access to Google Earth; pen or pencil and paper

Instructional Procedures

Introduction: Begin with a few questions about where their drinking water comes from. When they wash something down the drain, where does it go? If you put a little model boat on a local stream, what path would it take to get to the ocean? What is the closest ocean?

Tell the story of "Walter the Water Molecule."

Activity: After the story is over, discuss some of the same questions. Have them turn to a partner and discuss their watershed and take turns making up a new version of Walter.

Pass out "The Water Cycle" work sheet and, using pictures or short phrases, have them fill in the sheet as much as they can in just 5–10 minutes.

With access to Google Earth, ask students to zoom in and zoom out on their local watershed. Ask them to create several diagrams detailing individual steps in the water cycle. Following are a few examples:

- Using a world map image, students can trace global wind patterns, jet streams, and ocean currents; the path of water from the sea to a cloud to an aquifer outside of the United States. The students can trace weather patterns, emphasizing local weather and the path water takes from the sea to a cloud, to an aquifer to their home and back to the sea.

- Using a view of their national map, students trace local weather patterns and the path of water from the ocean (which would appear at the edge of the map); to the clouds; to the river, lake, or aquifer that provides drinking water for their city; to their city (sometimes the water source is quite far from the city); and back to the sea.

- With a map of their city, students map their local water supply. Have students draw a line from the treatment plant to their home, realizing that most water distribution pipes follow streets. Students create a diagram of their house and then map their home's plumbing. Encourage students to depict the pipes and outlets both outside and inside the house. Outdoor plumbing might include the water meter, spigots, and a sprinkler system. Also, be sure students include the wastewater plumbing, showing how water returns to the sewer system.

- With a diagram of the body, students map the circulatory system, showing the flow of blood between the heart and lungs as well as to other parts of the body. Encourage students to use different colors or arrows to indicate the flow of blood to and from the lungs. (Suggestion: Air leaving the lungs is oxygenated, so it is often depicted in red; blood that has made its rounds is depleted of oxygen, so it is often depicted as blue.)

After these maps and diagrams are created, students collate and compare them. This provides a basis for a discussion of macro versus micro systems. This is also great raw material for telling and writing their own stories about Walter or Wendy the water drop!

Assessment: Ask students to draw four diagrams: the world / their nation, their city water supply, their house's plumbing, their body and circulatory system. Each of these can be collected and graded for accuracy and detail.

Follow-Up Activities: These maps can be used as the outline for their stories in the following lesson plans.

The Circular Storytelling Game

This activity is described in detail on page 20. Students work in small groups to first tell and then write their own versions of the water cycle story. After working in small groups, ask students to work independently on their individual version of the story. If a picture is worth a thousand words, then their four maps and diagrams from the previous lesson can be worth four thousand words!

How Much of Me Is Water?

❖**Grade Levels**: 3–12   **Time estimate**: 30–45 minutes

❖**Science skills**: Metric Measurement; Prediction

❖**Objectives**: Students will make predictions about their water weight and exercise metric measurement skills and division/multiplication skills in computing this data.

National Standards

Science Standards

NAS 1 Science as Inquiry: Abilities necessary to do scientific inquiry; Understandings about scientific inquiry.

NAS 2 Physical Science: Properties and changes of properties in matter; Motions and forces; Transfer of energy.

NAS 3 Life Science: Structure and function in living systems; Reproduction and heredity; Regulation and behavior; Populations and ecosystems; Diversity and adaptations of organisms.

NAS 6 Science in Personal and Social Perspectives: Personal health; Populations, resources, and environments; Natural hazards; Risks and benefits; Science and technology in society.

Language Arts Standards

NCTE 7 Students conduct research on issues and interests by generating ideas and questions, and by posing problems. They gather, evaluate, and synthesize data from a variety of sources (e.g., print and nonprint texts, artifacts, people) to communicate their discoveries in ways that suit their purpose and audience.

NCTE 8 Students use a variety of technological and information resources (e.g., libraries, databases, computer networks, video) to gather and synthesize information and to create and communicate knowledge.

❖**Materials**: Bathroom scale; How Much of Me Is Water? work sheet (page 42; buckets; water; sand or hair, leather, and bones; pen or pencil and paper

---

Instructional Procedures

Introduction: Ask students to guess what percentage of their body is water. If they weigh X number of pounds, how many pounds or gallons of water do they hold? Knowing that science is all about metric measurement, ask how many kilograms they weigh and how many liters of water they hold.

Activity: As a simple math exercise, students weigh themselves and then compute how much of their weight is water and how much is solid. To calculate their water weight, students multiply their body weight by .60. To calculate their solid weight, students multiply their body weight by .40. (Humans are about 60 percent water by weight, 75 percent by volume.)

Students then calculate approximately how many gallons of water their bodies contain by dividing their water weight by 8.34. (Water weighs about 8.34 pounds per gallon.)

Once they have used the American Standard measurements of gallons and pounds, challenge them to convert these numbers to liters and kilograms. Science Made Simple has online calculators to help: http://www.sciencemadesimple.com/weight_conversion.php and http://www.sciencemadesimple.net/volume.php.

Each body contains varying degrees of fat, muscle, and bone, so students should be advised that their calculations will give them only a rough estimate.

To make this exercise more dramatic, create a graphic display. Line up four buckets at the front of the classroom. Fill one bucket with bones, hair, and leather. (Sand and dirt will work fine if this seems too macabre.) Fill the other three buckets with water.

As a simple transitional activity, brainstorm with the class a list of living things that contain water. Each student chooses one object and finds out how much of it is water. Each student produces a simple graph that demonstrates that the object is so much water and so much solid. For example, an elephant is 70 percent water, an earthworm is 80 percent, a jonquil is 65 percent, and a carrot is 90 percent.

Assessment: The work sheet can be collected and graded for both math and science.

Follow-Up Activities: This lesson is a lead-in to "How Much of This Is Water?"

How Much of Me Is Water?

I weigh _____ pounds.

You are 60% water by weight.

To find out how much of you is water, multiply your weight by .60, like this:

I weigh _____ pounds.

x .60

The water in me weighs _____ pounds.

A gallon of water weighs 8.34 pounds.

To find out how many gallons of water are in you, divide the weight of the water in you by 8.34, like this:

The water in me weighs _____ pounds.
divide by 8.34 pounds per gallon
My body contains _____ gallons of water!

*On the back of this sheet show the math for converting this
to kilograms and liters!*

How Much of This Is Water?

❖**Grade Levels**: 2–12   **Time estimate**: 45–60 minutes.

❖**Science Skills**: Metric Measurement; Classification; Communication; Identify Variables; Design Investigations

❖**Objectives**: Students will make predictions about the water weight of a variety of foods and exercise metric measurement skills and division/ multiplication skills in computing this data.

National Standards

Science Standards

NAS 1 Science as Inquiry: Abilities necessary to do scientific inquiry; Understandings about scientific inquiry.

NAS 2 Physical Science: Properties and changes of properties in matter; Motions and forces; Transfer of energy.

NAS 3 Life Science: Structure and function in living systems; Reproduction and heredity; Regulation and behavior; Populations and ecosystems; Diversity and adaptations of organisms.

NAS 6 Science in Personal and Social Perspectives: Personal health; Populations, resources, and environments; Natural hazards; Risks and benefits; Science and technology in society.

Language Arts Standards

NCTE 7 Students conduct research on issues and interests by generating ideas and questions, and by posing problems. They gather, evaluate, and synthesize data from a variety of sources (e.g., print and nonprint texts, artifacts, people) to communicate their discoveries in ways that suit their purpose and audience.

NCTE 8 Students use a variety of technological and information resources (e.g., libraries, databases, computer networks, video) to gather and synthesize information and to create and communicate knowledge.

❖**Materials**: Fresh and dried foods, such as jerky, bananas, apples, carrots, onions, green beans, and blueberries; pen or pencil and paper; metric scales

Instructional Procedures

   Introduction: Begin with a discussion of the importance of water: water is life. Extrapolating the last lesson, discuss how much of each living thing is made of water.

Activity: Students measure the amount of water in various foods. This is done as an experiment designed by the students. Ask them to determine how they might figure this out.

Begin by discussing dehydration. Show students several examples of dried foods, such as jerky or fruit leather. Then discuss characteristics that will help the students predict the water content of each type of fruit. For example, if a fruit is moist or juicy, it probably has a higher water content than a fruit that is dry or hard.

Working individually or in small groups, students choose one or more food items. After weighing the food item, students predict what percentage of the food item is water. The students record that percentage, then multiply the weight of the food item by the percentage figure to calculate the water weight of that food item. (Example: If a slice of banana weighs 20 grams, and the students predict that 20 percent of that weight is water, then the calculation is: 20 grams x .20 = 4 grams.)

Ask students to design a method to test their prediction. They will need help implementing their plans.

If you have access to an oven at school, dehydrate the food by spreading it in a thin layer on a baking sheet and baking it at 200 degrees or lower until dry. If you do not have an oven at school, either ask students to dehydrate the fruit at home or do it for them. A quick online search found several designs for solar dehydrators, but probably the easiest to build is here at J. R. Whipple: http://www.jrwhipple.com/sr/soldehydrate.html. Or, check your library for plans to construct a simple solar dehydrator.

Students weigh the dehydrated food items and compare the actual weight to their estimate. This can be expressed in bar graphs or pie graphs.

Assessment: Students can be graded on the design and implementation of their inquiry, the accuracy of their prediction, and the mathematics of their measurements, percentages, beginning and final weights.

Follow-Up Activities: Students could dehydrate a wide range of fruits and vegetables. As a class you could share lunchtime snacks made from wholesome and nutritious dried fruits.

Evaporation

❖**Grade Levels**: K–5

Time estimate: 45 minutes the first day, and then a few minutes each day for several days.

❖**Science Skills**: Metric Measurement; Prediction; Design Investigations

❖**Objectives**: Students will design an investigation to test various effects on evaporation.

National Standards

Science Standards

NAS 1 Science as Inquiry: Abilities necessary to do scientific inquiry; Understandings about scientific inquiry.

NAS 2 Physical Science: Properties and changes of properties in matter; Motions and forces; Transfer of energy.

NAS 3 Life Science: Structure and function in living systems; Reproduction and heredity; Regulation and behavior; Populations and ecosystems; Diversity and adaptations of organisms.

NAS 6 Science in Personal and Social Perspectives: Personal health; Populations, resources, and environments; Natural hazards; Risks and benefits; Science and technology in society.

Language Arts Standards

NCTE 7 Students conduct research on issues and interests by generating ideas and questions, and by posing problems. They gather, evaluate, and synthesize data from a variety of sources (e.g., print and nonprint texts, artifacts, people) to communicate their discoveries in ways that suit their purpose and audience.

NCTE 8 Students use a variety of technological and information resources (e.g., libraries, databases, computer networks, video) to gather and synthesize information and to create and communicate knowledge.

❖**Materials**: Glass jars; water; salt; food coloring; a flower or leafy branch; microscope; slide showing stomata; barometer; thermometer; hygrometer

Instructional Procedures

Introduction: We have heard the word *evaporation*, but what does it mean and how does it work? Through Socratic questions discuss evaporation and encourage students to ask good scientific questions.

Activity: Students design experiments with evaporation.

Collect about two or three dozen glass jars of uniform shape, at least two for each student. Include one jar with an especially wide mouth and one with a narrow mouth. The jars should hold about 1/2 liter of water.

Pour 500 mL of water into each jar. Mark the water level on the outside of each jar.

Place one jar in a sunny place. Seal the other jar and set it aside. These two jars are your control group. The first is the standard for comparing the speed of evaporation from the various jars. The sealed jar shows how much water you started with.

The rest of the jars are used to test a wide range of influences upon evaporation. Allow the students to test various influences. Here are a few to start the conversation:

- What might happen if we add a tablespoon of salt to one jar?

- What if we place one jar in a cool, dark corner?

- What if we add black dye or dark food coloring to another jar?

- Use one jar with a very wide mouth and one jar with a very narrow mouth.

- Place a fresh flower in one jar. This measures the effects of evapo-transpiration as well as evaporation. To help explain the concept of evapo-transpiration, obtain a slide of stoma and allow students to examine it under a microscope.

In addition to altering the water in the jar, consider outside influences on evaporation. For example, students chart the temperature, barometric pressure, and relative humidity each day to monitor their effects on evaporation. (These numbers are easily obtained online at WeatherChannel.com or on your smart phone, if you do not have the tools to measure them yourself.)

What are some other options? Allow students to suggest research proposals. Each day, students mark the water level in each jar and make a corresponding mark in a journal or on a chart they design. Students predict which jar will empty first, second, and last. After they collect the data they interpret the results and write a brief report emphasizing which variables had the greatest effect on the rate of evaporation.

Assessment: Students should be encouraged to keep a journal throughout the process with daily reflections, predictions, and explications of what is going on. These journals can be evaluated with an eye toward encouraging inquiry. They can be graded on the design and implementation of their inquiry, the accuracy of their prediction, and the mathematics of their measurements. Their final reports can also be collected for a grade.

Follow-Up Activities: Encourage students to follow their own lines of inquiry, to ask what, if, and why questions, and more importantly to look for their own answers!

The Water Cycle

Use a pen or a pencil to trace the possible routes of Walter through the water cycle. Add detail to this very basic beginning of an illustration. Remember, the basic steps are:

- evaporation, when water becomes air
- condensation, when the water drops come together to from dew, clouds, or fog
- precipitation, when water falls as snow, sleet, hail, or rain
- collection, when water gathers in a lake, pond, aquifer, or body

Use the picture below and the story of "Walter the Water Molecule" as a starting point for your own version of the story of the water cycle. You could start in a lake, ocean, or river, rise into the air (evaporation), form a cloud (condensation), float over the land, and fall as sleet, snow, or rain (precipitation). Flow through a river, plants, animals, and into a human (collection), then back into the air, a cloud, and into the ocean. Add pictures of your own to this drawing. The more details you add, the better your story will be.

From *Learning from the Land: Teaching Ecology through Stories and Activities,* Second Edition by Brian "Fox" Ellis. Santa Barbara, CA: Libraries Unlimited. Copyright © 2012.

A Story Map for "Walter the Water Molecule"

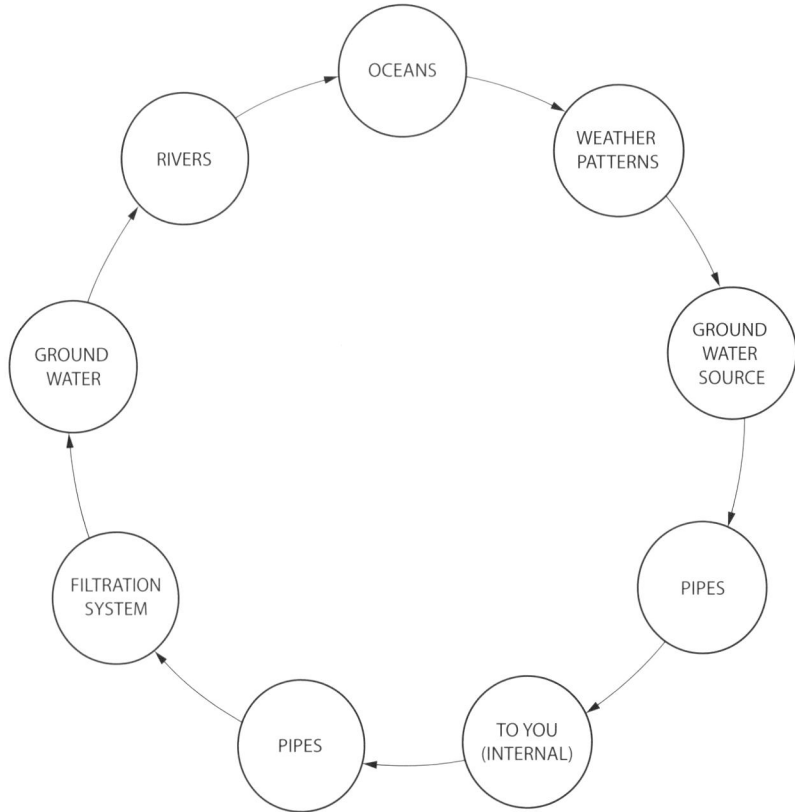

After listening to the story of "Walter the Water Molecule," create a flow chart that tracks your water from the ocean to you and back to the ocean again! The key words are evaporation, condensation, and precipitation.

Look at a map of the world and study weather patterns to figure out where your clouds gather moisture.

Look at a map of your state or county and trace the creeks and rivers that flow through your neighborhood. (Include the other paths of water through plants and animals that share your water supply.)

Study the groundwater underneath your home. Look for maps of aquifers and underwater streams and lakes.

Where does your drinking water come from? Trace a map of your city or county and map the path of water from the filtration plant to your home.

Draw a diagram of your house and map the pipes to your faucet.

Draw a diagram of your body and map the path of water through your bloodstreams.

Now reverse the process and, step by step, trace the water from you through the filtration process, rivers or streams, back to the ocean.

Use this information to write your own exciting adventures of "Walter the Water Molecule." Rewrite and edit your story. Draw pictures and publish your own book. Send me a copy and I may use your story on my Web page!

From *Learning from the Land: Teaching Ecology through Stories and Activities,* Second Edition by Brian "Fox" Ellis. Santa Barbara, CA: Libraries Unlimited. Copyright © 2012.

The Web: Lessons in the Web of Life

May I be worthy of my meat.

—Wendell Berry, from "Prayer After Eating"
in *The Country of Marriage*

Comments to the Teacher

In PART, THIS STORY WAS created in response to children at a summer camp complaining about mosquitoes. I would tease them, saying, "Do you like listening to the birds sing? Do you like eating the fish we catch? Then you'd better feed the mosquitoes, because they feed the birds and fish!" This evolved into a story that celebrates all the creatures of a wetland ecosystem. It is about the food web and our place in it.

When you tell this story, adapt the particulars to fit your life experience. Most storytellers automatically adapt existing stories in this way. It's like tailoring a suit; it feels more comfortable if it fits you. Name and describe your favorite fishing hole. Change the species of the bird, frog, and fish to animals that live in your ecosystem. Mosquitoes and dragonflies are found in most parts of North America, but the heron could be an egret, the bullfrog could be a leopard frog, and the bass could become a tarpon. If you never went to camp or worked at one, you could change the setting to a river, farm pond, or lake you have visited. Lars could become one of your children or your father or uncle—he is the person you are most likely to go fishing with.

If you have never gone fishing, fake it! There is enough detail in this story to be convincing. If that feels uncomfortable, you could say you heard the story from an avid angler, and you could change the first-person pronouns to third-person pronouns. But this story works best told in the first person. The most important goal of this tale is to make personal and tangible the abstract concept of the food web. In all honesty, if you have ever been bitten by a mosquito or eaten some food, then you are a character in this tale!

And now the story . . .

This story takes place at a camp in Michigan called Camp Niobe. The word *niobe* is Greek for willow. The lake at that camp was surrounded by beautiful weeping willow trees, so you can see where the camp got its name. Something is going on at Camp Niobe, something that is very obvious. As a matter of fact, it's so obvious that no one even notices. But it's very special. By listening to this story, maybe you can figure out what is going on.

Like a jigsaw puzzle or a quilt, this story is made up of many small pieces. At first the pieces don't seem to go together, but if you listen carefully you will find the pieces fit perfectly. Your job is to try to fit them together, to solve the mystery that is right before your eyes. How do these pieces go together?

Here's the first piece:

In one of the cabins at Camp Niobe, in the rafters above my bed, of all places, there lived a family of raccoons. If you know anything about raccoons, you know they make a terrible mess. Gosh, it really smelled in there! The cabin was the raccoon family's winter home, but during the summer they moved out to a hollow log behind the cabin; that was their summer home. Then, during the summer, the cabin became the campers' home. That cabin was like a time-share condo!

I was a counselor at Camp Niobe. I had studied with several Native American elders who taught me about wild edibles and herbal remedies. Most summers I came to Camp Niobe early, before any of the campers arrived. My job was to clean out all the cabins, but I really came because I wanted to see how well I would do if I just lived off the land. I was all on my own, out in the wilds of Michigan, with only the earth to rely on.

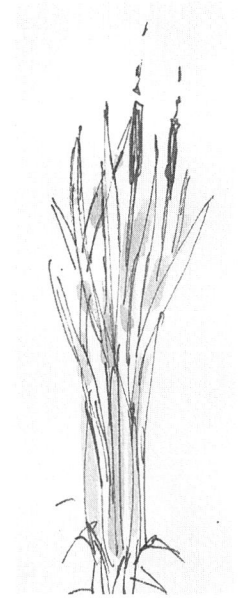

I brought very little food to the camp, just some brown rice and some seeds. I sprouted some of the seeds—have you ever eaten sprouts? —and I planted the rest of the seeds in a garden. I gathered wild greens from the fields. I went fishing for dinner and usually caught a nice bass or perch or bluegill. Sometimes I would dig up cattail roots. That's right. Cattail roots, if you cook them right, are a little bit like potatoes.

I found plenty to eat. Sometimes I caught frogs and ate frog legs. You are probably saying "Yuck!" But have you ever had frog legs? Don't knock it until you try it! They are delicious. Some people say that they taste like chicken, but they're much better than chicken if you sauté them in a little butter and garlic. I found wild garlic growing at Camp Niobe.

And that wasn't all. One field was filled with dandelions. Early in the spring, dandelion greens are quite delicious; late in the year they become bitter. I gathered wild greens and made a delicious salad with wild violet flowers for flavor and color.

Now this is a good time to bring up an important point; I could gather wild foods because I had studied with an expert who taught me what was safe to eat. Many wild plants are poisonous. So, if you decide to try this out, don't just start grabbing any old weeds and eating them, because you might eat something poisonous. Go with an expert. Know what you're looking for.

After I made my dinner, I would sit on a picnic table right outside the dining hall. Before me was the meadow where I had just gathered my greens. While I was eating my dinner of frog legs or fish and salad and roots, a family of groundhogs would come to play in the field. There was a big mama groundhog with her three babies, rolling and wrestling and playing. Every night they entertained me while I ate my dinner.

As I had learned from the elders, if you walk without making any sounds, then you can get really close to wild animals. I had practiced walking without making a sound, and I could get very close to these groundhogs. If the mother groundhog looked at me, I would freeze. But as soon as she looked away, I would take another step. In that way I could get close, within about 30 feet or so. If I got too close, the mama groundhog would stare at me, and I would freeze. If mama groundhog looked away, I might try taking another step, but if I got too close—[Whistle!] —that mama groundhog would whistle. The babies would disappear into their hole, and mama would stand guard at the entrance.

Raccoons, greens, roots, frogs, and groundhogs. We already have five pieces of this story. Like I said, it's like a jigsaw puzzle. Your job is to figure out how the pieces go together.

Here's another piece:

One of my favorite things to do at Camp Niobe was fishing. I loved to go fishing early in the morning and late in the evening. If you know anything about fishing, you know that sunrise and sunset are the best times to be on the water. I knew that and so did Lars, the camp's maintenance man.

After the camp opened for the summer, I was not alone. Sometimes Lars woke me up at four o'clock in the morning. Lars would tiptoe into the cabin, where some boys and I were sleeping. Lars would whisper, "Come on! Get up! Come on!"

I would say, "Get out of here, Lars! I want to sleep!"

But Lars knew better. I didn't really want to sleep; I wanted to fish. So Lars would keep whispering until I was awake, "No, no! You said you wanted to go. Come on, let's go."

Quietly, so as not to wake the campers who shared the cabin, I would grab my tackle box and fishing pole. Then Lars and I would tiptoe out of the cabin. *Screeeech!* Of course the door squeaked!

"Lars, you've got to oil that door," I would tease. Lars was the maintenance man, after all.

One early morning, the clouds were so thick you couldn't see the stars or the moon. Lars and I climbed into the boat and started rowing. The lake was beautiful, with weeping willow trees lining the shore and fog rising off the water.

Fog rises off the lake in the late summer and fall, but not so much during the spring. Do you know why? Let's think about this. When you set a cold glass of water outside on a hot day, what happens? The outside of the glass gets wet, right? Is that water leaking out of the glass? No, no! It's moisture from the air trying to warm up the cold glass. The air is

warm and the water is cold. That action of moisture in the warm air trying to warm up the cold glass happens when unequal forces act on one another, trying to reach equilibrium. It is part of the natural process of thermal dynamics.

Now, here's something interesting: Even as the warm air is trying to warm up the cold water in the glass, the cold water in the glass is trying to cool off the air. The water does that by evaporating.

Think of the lake as a huge glass of water. But this water is warm. Every day the sun beats down on the water and slowly, over the course of a long, hot summer, the lake warms up. Then in late summer, the nights cool off. On some mornings the air is quite chilly. When that happens, the lake, being warmer than the air, gives off moisture to try to warm up the air. That evaporation makes fog.

Back to our foggy morning. The fog—moisture rising off the lake—was so thick I could hardly see Lars in the back of the boat. The fog was so thick I couldn't really see where I was rowing.

Suddenly, *caw! caw! caw!* A great blue heron took flight right next to the boat! Its wings nearly beat us over our heads, and its feet dangled in our faces as it disappeared into the fog.

What do you know about the great blue heron? It's one of my favorite birds. Great blue herons stand about four or five feet tall. They have long, thin legs and a long, skinny neck and a long, yellow beak. Their wings are longer than my arms.

If you've never seen a great blue heron hunt, you've missed a pretty sight. They fold their wings back as they wade through the shallow water looking for fish or frogs. They move gracefully, like ballerinas. They move slowly, standing on one leg, moving forward, standing on the other leg. Sometimes, if you stare at a great blue heron while it is hunting, you will not even see it move. That's how slowly they move. Of course, that's the point: they move so slowly that the frogs and fish don't see them coming. [*Impersonate a heron in a heron dance!*]

Just as the great blue heron stalked its prey, I sometimes stalked the great blue heron. I didn't want to eat the bird— I just wanted to watch it fish! I would swim out across the lake like a frog or an alligator, with just my nose and eyes out of the water. That way, I could get close to the great blue heron, just a few yards away. Moving slowly and quietly, I would get as close as I dared to the bird's sharp, pointed beak. One time, when I was stalking a heron, I saw the bird freeze and stare at the water. Slowly it pulled its head back, cocked its neck, and *splash!* Into the water it stuck its beak and brought up a fish.

Whether it catches a frog or a fish, the great blue heron has an unusual way of eating. It turns the frog or fish around in its long, skinny beak and swallows it head first. Why does the heron do that? Why doesn't it swallow its prey backward or sideways? It never did; the heron always turned its prey around and swallowed it head first.

Let's formulate a hypothesis. Do you know what a hypothesis is? It's an educated guess. Actually, it's more than a guess, more of a prediction based on the facts at hand. What do you think? Why do you think the heron swallows its prey head first? Take a moment and think about the facts. What do you know about herons and fish? [*Discuss this with students. Ask them to formulate a hypothesis with a partner. Give them a minute, then call on a few students to share their hypotheses.*]

My hypothesis is this: If the heron tried to swallow the fish tail first, the fins would open up and the fish would get stuck in the heron's long, skinny throat. If the heron tried to swallow the fish sideways, it wouldn't fit. But when the heron turns the fish around and swallows it head first, then the fins close and the fish slides right down the heron's long, skinny neck. Its neck is so thin that sometimes you can see the outline of the fish as the heron swallows it.

Fog, fish, frogs, heron. Four more pieces of the story, You probably haven't figured out how these pieces fit together yet. But listen closely: they do.

Here's another piece of the story:
One hot summer day—one of those days when the sun is like a ball of fire— I was hanging out on the dock. It was so hot I was sweating, just sitting still. Soon my mind floated off like a little cloud in an otherwise blue sky. Do you like to daydream? As my mind drifted, I looked down into the water on the right side of the dock. Something was splashing around down there.

Now, the left side of the dock was the swimming area. Years ago, someone had dug out the lake bottom to make it deep for swimming. But on the right side of the dock the water was very shallow. It was covered with lily pads and duckweed and cattails. That's where I saw something splashing around. When I looked closer I noticed that something was a dragonfly nymph, squiggling in the swampy water.

Do you know what a dragonfly nymph is? It's an immature dragonfly. Do you know what it looks like? It looks like the Creature from the Black Lagoon. It's black and slimy, with huge mouth parts. These bugs can tell you how clean the water is! If you have water you have life. If it is heavily polluted then there is no life. If you find midges and water pennies, it is mildly polluted, but dragonfly, caddis fly, and damselfly larvae are a sign of clean water.

As I watched, the dragonfly nymph crawled up the side of a cattail. As it did, the back of its shell cracked. To my amazement, that ugly creature of the swamp climbed out of its shell and became a beautiful creature of the air.

Have you ever seen a green darner dragonfly? The large ones that are turquoise blue or green? They have four wings that stretch out. You can see through the wings, except for the black lace that holds them together. When they hold those wings just right into the sun, they

refract the light to cast a rainbow of colors. Right before my eyes, this dragonfly came out of its shell—it metamorphosed. The dragonfly pumped fluid into its wet crinkly wings, filling its wings until they stretched out before my eyes.

As its wings began to dry, the dragonfly told me the rest of this story, and all of the pieces began to come together.

Here is the final piece of the story:

One summer night, I was out in the early evening—about sunset—doing a little fishing with Lars, the maintenance man at camp. You know, sunrise and sunset are the best times to go fishing. But if you ever go out on the water in the early evening, you know it's the best time for something else, too! Bzzz! Mosquitoes everywhere! Ugh! Ow! Slap! Bzzz! There goes one with a pint of my blood! Bzzz! Flying slow! Bzzz! Its belly big and red with my blood! Bzzz!

Then, as I was watching a mosquito make off with a pint of my blood . . . *snatch!* A great big dragonfly grabbed that mosquito right out of the air. A huge dragonfly! Bzzz! With four big wings, bzzz! Flying around, eating hundreds of mosquitoes. Bzzz!

Did you know dragonflies eat mosquitoes, deerflies, horseflies, and all kinds of biting insects? Let's hear it for the dragonflies! Hip-hip-hooray! They are the falcons of the insect world. They are one of my favorite bugs!

But dragonflies do more than just eat mosquitoes. Sometimes one dragonfly meets another one. And they start talking about poetry and Shakespeare. And they feel a spark that bursts into a passionate flame. And they begin flying together,

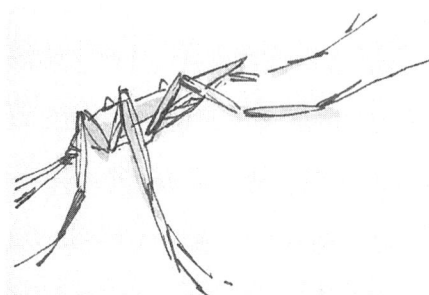

one dragonfly touching its tail to the back of the other as they fly along.

Then one dragonfly flies off by itself and lands on a lily pad. It pushes its tail underwater and begins to lay eggs. One egg and one egg is two. Two and two are four. Four and four are eight. Help me out here if you can. Eight and eight are . . . 16. Sixteen and 16 are 32. Thirty-two and 32 are? Sixty-four, right! And 64 and 64? One hundred and twenty-eight. [*Allow them to work out the math with you, the power of two, pausing for a second for someone to guess the next step.*] And then . . . 128 + 128 . . . 256 and 256 . . . 512 and 512 . . . 1,024. I am going through this math for two reasons: to give you an idea what a huge number 1,000 really is *and* because if you are going to be a scientist then you need math skills! Exercise your math muscle and you too can be a human calculator!

One dragonfly lays more than 1,000 eggs. Not all of them survive, but many of them do. They become nymphs, or larvae. One day, they break out of their shells and become dragonflies—just like the one that crawled up on the cattail next to the dock where I was daydreaming just a few days ago.

All insects go through these changes called metamorphosis. Some do a complete metamorphosis, making a dramatic change from egg to larvae to pupa to adult. Others make more gradual changes from egg to nymph to larger nymph to adult. But it's not only bugs that change as they grow. All living things go through transformations. You! You started out as an egg inside your mother's womb. And then you were born as a cute little larva, I mean, baby. And then

you grew to be an ugly pupa, or adolescent. Someday, you'll sprout wings and become an adult, as some of us already have.

But let's get back to the story. There I was, just about sunset, being eaten alive by mosquitoes when . . . *snatch!* A dragonfly ate a mosquito that was full of *my* blood. The dragonfly had no sooner grabbed that mosquito then . . . *fwumpp!* A big bullfrog whipped out its tongue and swallowed the dragonfly whole. It was a h—u-u-u-g-e bullfrog. The frog that sings late at night, down by the lake. Burrumph! Burrumph!

The frog lives in two worlds, which is what *amphibian* means. He swims in dark pools and leaps about upon the land at dusk. He sings to the moon, hoping another frog will hear his song, but mostly singing because it feels good in his throat, in his bulging throat. When he is bored or hungry, he caches little bugs with his sticky tongue. He is most happy sitting in the cool, muddy, shallow water, smelling the warm, muddy night air, and singing because it feels good in his throat, in his buuulllging throat.

After the bullfrog caught the dragonfly, it began to swim across the lake. Imagine that you are there now. The lake is as smooth as glass and as black as ink. Reflected in the water you can see all the stars. There's one star in the cup of the quarter moon. The little bit of moonlight is shimmering on the water. The only ripples are made by the bullfrog swimming. [*Make a loud SPLASH!*] Up comes a bass! A largemouth bass swallows the bullfrog whole. The bass swims off to the bottom of the lake, resting and digesting with the bullfrog in his belly.

Just a few days later, Lars, the maintenance man said, "Fox, how'd you like to do a little fishing?"

I said, "What kind of a question is that? Of course I would. Why don't we try the west end of the lake tonight? We haven't fished there for a while."

Lars said, "Sure, why not? You're rowing."

So that evening, we went out on the lake again. Imagine you are sitting in a boat in the middle of a lake at sunset. Can you see the sun setting? The clouds are orange, purple, pink, maroon, and magenta. It looks like the sky is on fire. In the glassy smooth reflection of the water you can see the sun setting twice, reflecting into itself.

I rowed across this lake on fire until we neared the far shore. The water near the far shore was choked with weeds: duckweed, lily pads, and cattails. But I noticed a spot as big around as a basketball hoop where the water was clear. There were no weeds there. I said, "Lars, I'll bet you there's something big hiding under there."

Lars replied, "So what if there is? You can't cast into that little spot from here."

I said, "Watch!" I tied on my favorite lure—a jointed minnow, hand-carved out of balsam wood and painted to look just like a real minnow. It has a little swivel in the middle of its back, and when you cast it, it floats on top of the water and when you wiggle it, it swims like a wounded minnow. That's good. If you know anything about predator-prey relationships, you know that predators always go for the weak or wounded prey. Bass, in particular, don't chase their prey. They hide and wait to *pounce* on their food, like a tiger waiting in the tall grass! So I tied on the lure and stared at that spot of clear water, concentrating, meditating: I think I can, I think I can, I think I can, I know I can, I know I can, I know I can. Bzzzzz, splash! The lure landed right on the far edge of the circle of clear water. I can! I can!

My father, who taught me everything I know about fishing, always said, "When you're casting a surface plug, you want to let your lure sit still for a while. You want the fish to get used to seeing it so they don't think it just fell out of the sky." When my lure landed, it sent out gentle ripples. I waited for the concentric circles to disappear. When the water was as smooth as glass, I reeled in the slack of my fishing line. I elbowed Lars, saying, "If there's a big one down there, I'm going to catch it."

Carefully I wiggled the lure so it swam like a wounded minnow. Nothing happened. Now the lure was right in the middle of that circle of clear water. The fish would have to bite the lure while it was in the clear water; once the lure got tangled in the weeds, it was all over.

Once more I waited until all the concentric circles disappeared. As I reeled in the slack, I knew it was now or never. I wiggled the lure again, making it swim like a wounded minnow. Nothing happened. The lure was in the weeds now, and I knew I wasn't going to catch anything. Disappointed, I started to reel in the lure—and *splash!* A bass took it. Sometimes they wait until that third time.

I set the hook hard. The fish dove off to the left and tried to tangle the line in the weeds. But I kept tension on the line. I was reeling it in when . . . *splash!*

That fish exploded out of the water. Have you ever seen a fish dance across the surface of a lake? It is trying to spit out the hook. I wouldn't let it. I kept the tension on the line and reeled it in.

I pulled up the fish, worked my fingers in under its gills. Wow! I had never caught a largemouth bass so big. It was about 20 inches long and weighed about 5 pounds. Its mouth was so big I could fit my fist into it without touching one of its tiny little teeth.

Now, usually if you catch a big fish, it's a good thing to let it go. That helps to keep a healthy balance between predator and prey. But I hadn't had any dinner. In this case, I was a predator, too.

So I took the fish and rowed back across the lake. I took the fish up to the picnic table outside the dining hall.

Do you know what *déjà vu* is? It is when you are someplace, doing something, and you feel like you've done it before. I sat on the picnic table and looked out across the field. Earlier in the summer, I had watched a mama groundhog and three little groundhogs wrestling and playing; now I saw a big-mama counselor and three little campers rolling and wrestling and playing. The counselor blew her whistle. Some of the campers, or the groundhogs or whatever, went off to their hole—I mean, their cabin.

My campers gathered around the picnic table. "What'd you catch? What'd you catch?"

I showed them the fish. I decided to take advantage of the teachable moment and do a little dissection lesson.

Imagine with me. See how the top of the fish is dark green and black, and the bottom is white? That is perfect camouflage for the fish's habitat. Imagine that you are floating on the surface of a lake, looking down. What do you see? Green weeds and black mud, and the top of the fish is dark green with black spots. Now imagine you are swimming along the bottom of the lake and you look up through the water. What do you see? The sun sparkling and the clouds, light colors, and the belly of the fish is white. In its environment, it has perfect camouflage.

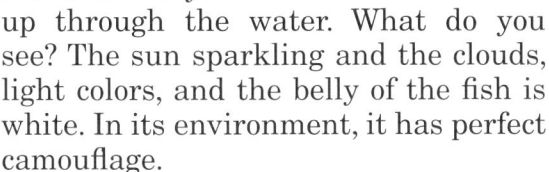

Now imagine how a fish breathes. When its mouth is closed, the gills open. When its gills are closed, the mouth opens. You've all seen fish lips, right? [*Demonstrate*.] That's how a fish breathes. It breathes water and pumps that water across its gills. Everyone, take a deep breath. We pump air into our lungs; fish pump water across their gills, and that's how a fish breathes.

Now, let's take a look at the belly.

Imagine cutting the fish open. The fish's belly is full. Let's see what it ate. We'll cut the stomach open and . . . *a-ha!* a bullfrog! There was a bullfrog in the stomach of that fish!

What do you think might be in the stomach of that frog? A dragonfly? And what do you think might be in the stomach of that dragonfly? A mosquito? A *mosquito*? And I thought about the mosquito that had bitten me just the other day. Bzzz! Bzzz! Slap!

I call this story "The Web" for the web of life that ties us all together. Now do you see how all the pieces fit together?

Discussion Topics

The first question is simply this: Who eats whom? Before passing out the Adventures in the Web of Life work sheet, discuss the aquatic food web illustrated in this story.

How do the pieces of the story fit together? As a class discuss each part of the food web illustrated in this story and how each creature is eaten by or eats a few other creatures. Be sure to tie in some of the stray characters. What about the great blue heron? The raccoon? What about the groundhog? How do these creatures tie into the web of life?

You may also want to raise issues related to disruptions in the food web. What if you eliminated the largemouth bass or the large predators from the lake?

What if you poisoned the lake with pesticides in a well-intentioned effort to control mosquitoes? If a dragonfly ate 100 poisoned mosquitoes, a frog ate 10 dragonflies, a fish ate 5 frogs and 5 dragonflies, and you ate 3 fish? You would eat 16,500 particles of pesticide $([100 \times 10 \times 5] + [100 \times 5]) \times 3 = 16,500$). This process is called biological magnification; it explains why the pesticide DDT decimated bald eagles during the 1960s.

What are other threats to the balance in nature? What happens when an alien plant is introduced, and its vigorous growth decreases the sunlight and nutrients available for native plants? What if an alien fish is introduced and eats the plankton that is the foundation of this aquatic food web? What can we do to protect or promote the balance of ecosystems? State departments of fish and wildlife monitor the fish populations of many public and private ponds and lakes. Like fish, many insects (including their nymphs) are indicator species; by keeping an eye on the indicator species you can discover problems before they explode. What else can we do to protect ecosystems?

Adventures in the Web of Life

❖**Grade Levels**: K–12   **Time estimate**: 50 Minutes

❖**Objectives**: Students will first take apart the story of "The Web" and then use it as a model for writing their own story.

They will demonstrate an understanding of the food web and their role in it, so it is no longer an abstract idea, but what they had for lunch.

❖**Science skills**: Observation; Classification; Communication; Prediction; Inference; Identify Variables; Formulate Hypotheses; Reorder, Analyze, and Draw Conclusions; Design Investigations

National Standards

Science Standards

NAS 1 Science as Inquiry: Abilities necessary to do scientific inquiry; Understandings about scientific inquiry.

NAS 3 Life Science: Structure and function in living systems; Reproduction and heredity; Regulation and behavior; Populations and ecosystems; Diversity and adaptations of organisms.

NAS 6 Science in Personal and Social Perspectives: Personal health; Populations, resources, and environments; Natural hazards; Risks and benefits; Science and technology in society.

Language Arts Standards

NCTE 1 Students read a wide range of print and nonprint texts to build an understanding of texts, of themselves, and of the cultures of the United States and the world; to acquire new information; to respond to the needs and demands of society and the workplace; and for personal fulfillment. Among these texts are fiction and nonfiction, classic and contemporary works.

NCTE 4 Students adjust their use of spoken, written, and visual language (e.g., conventions, style, vocabulary) to communicate effectively with a variety of audiences and for different purposes.

NCTE 5 Students employ a wide range of strategies as they write and use different writing process elements appropriately to communicate with different audiences for a variety of purposes.

NCTE 7 Students conduct research on issues and interests by generating ideas and questions, and by posing problems. They gather, evaluate, and synthesize data from a variety of sources (e.g., print and nonprint texts, artifacts, people) to communicate their discoveries in ways that suit their purpose and audience.

NCTE 8 Students use a variety of technological and information resources (e.g., libraries, databases, computer networks, video) to gather and synthesize information and to create and communicate knowledge.

NCTE 11 Students participate as knowledgeable, reflective, creative, and critical members of a variety of literacy communities.

NCTE 12 Students use spoken, written, and visual language to accomplish their own purposes (e.g., for learning, enjoyment, persuasion, and the exchange of information).

❖*Materials*:

Adventures in the Web of Life work sheet (page 63)

Exploring New Environments work sheet (page 64)

Crayons or markers; pen or pencil and paper

Access to library research materials and the Internet

Instructional Procedures

Introduction: Introduce the story with the idea that as they listen, they are like detectives trying to solve a mystery, trying to figure out how the pieces go together.

Tell the story of "The Web"!

Activity: Afterwards lead a brief discussion of who eats whom. What are the parts of the food web and how are we part of that web?

Pass out the work sheet, Adventures in the Web of Life. (To save paper, print the Exploring New Environments work sheet on the back.) Allow students just a few minutes to work individually, drawing the lines that connect predator to prey. Ask each student to choose a partner and to share their food webs as an opportunity to learn from each other and to learn about each other's unusual eating habits. Have you ever eaten a groundhog?

Ask students to fill in the Exploring New Environments work sheet. Using the instructions on the work sheet, challenge students to choose an environment they know something about. Ask them to list and then draw several of the plants and animals that live in that environment. Using a writing process approach, guide students through the construction of a new version of "The Web." They could first write one or two descriptive sentences about each character, followed by two or three sentences about the setting or habitat. Students could make a story map or outline of their plots. These rough notes could then be rewritten into story form.

Give students just a few minutes to turn to a partner and take turns telling, not reading, their new story. For homework, challenge students to surf the web to learn more about their chosen ecosystem and the plants and animals who live there. With careful rewriting and editing, students could polish a final draft.

Assessment: The work sheets can be collected and graded based on the number of lines correctly drawn and the number of plants and animals included on the back.

Their stories can be assessed based on both the science content and the creative writing skills, so students could receive two A's for one paper: an A in science for good descriptive writing and details about their chosen habitat and the diversity of species; and an A in language arts for spelling, grammar, punctuation, and the stylistic qualities of flow, detail, vocabulary, and so forth.

Follow-Up Activities: These stories could then be illustrated and collected into anthologies based on the environments described, for example, Stories from the Tundra or Stories from the Rain Forest. Students could practice and then perform their stories for their class or other classes. Several students who chose the same ecosystem could combine their stories and develop a skit based on their combined story. (For an early elementary version of this skit please see "Song of Life" on page 82.) Small groups could also develop a food web diagram to display on a bulletin board along with their stories.

Adventures in the Web of Life

Draw the food web, connecting animals to their sources of food. Who eats whom? Remember that many predators eat more than one kind of prey, and many animals are prey for more than one kind of predator!

Use this web as a model to write your own story about your adventures in the web of life! Take me out of the story and place yourself in it. Maybe instead of a bass, what kind of fish do you eat? Who eats you?

From *Learning from the Land: Teaching Ecology through Stories and Activities,* Second Edition by Brian "Fox" Ellis. Santa Barbara, CA: Libraries Unlimited. Copyright © 2012.

Exploring New Environments

Imagine that you are in a particular environment: a rain forest, desert, prairie, grasslands, coral reef, taiga or boreal forest, tundra, city park, or any other environment you can imagine. List the plants and animals that live there. Draw a picture of each plant and animal in the list. Now think about who eats whom. Write a story about this food web in this environment. Put yourself in the story. What might you eat? What might eat you?

Tell your story to a partner and then listen to your partner's story. Rewrite, edit, and type your story to share with your class.

Plants and Animals Drawings

Write your list here.   Draw your plants and animals here.

_____

_____

_____

_____

_____

_____

Making a Food Web

❖**Grade Levels**: K–5   **Time estimate**: 30 minutes

❖**Science skills**: Observation; Classification; Prediction; Inference; Identify Variables

❖**Objectives**: Students will gain a visceral, kinetic understanding of the food web.

They will practice asking good questions and strengthen their use of deductive logic as they discover which plant or animal they represent.

National Standards

Science Standards

NAS 1 Science as Inquiry: Abilities necessary to do scientific inquiry; Understandings about scientific inquiry.

NAS 3 Life Science: Structure and function in living systems; Reproduction and heredity; Regulation and behavior; Populations and ecosystems; Diversity and adaptations of organisms.

NAS 6 Science in Personal and Social Perspectives: Personal health; Populations, resources, and environments; Natural hazards; Risks and benefits; Science and technology in society.

Language Arts Standards

NCTE 4 Students adjust their use of spoken, written, and visual language (e.g., conventions, style, vocabulary) to communicate effectively with a variety of audiences and for different purposes.

NCTE 11 Students participate as knowledgeable, reflective, creative, and critical members of a variety of literacy communities.

❖**Materials:** Ball of string; index cards; markers; tape, clothespins, or safety pins; paper and pencil

Instructional Procedures

Introduction: To complete this activity, students must understand not only predator–prey relationships but also the cycle of life. If you have not already performed "The Ballad of Rusty and Nancy" for your class, take a few moments to explain life cycles, including the fact that when animals die they rot and return to the soil, which is the source of nutrients for plants. Thus, even plants belong to the food web. In addition, you might introduce the terms *carnivore, omnivore, herbivore*, and *detritivore*. (Detritivores are carrion eaters.)

Activity: With a ball of string and some index cards, students create a food web in the classroom.

Count out enough index cards to provide one to each student in the class. On each index card write the name or paste a picture of a plant or animal. Be sure to include a good balance of food plants and animals, producers and consumers, predators and prey.

To do the activity with the whole class, create one set of cards with plants and animals from a single ecosystem. To do this with several small groups, make several sets of cards, each set devoted to a different ecosystem, such as rain forest, savanna, tundra, wetlands, and so forth. This also works well with a set of wildlife and wildflower postcards.

Using tape, clothespins, or safety pins, attach one card to the back of each student. Do not allow the student to see the card. Then tell the students to figure out what kind of animal or plant they are by asking each other yes-or-no questions. Questions could refer to size and color or to habitat and diet. For example, am I brown? Am I bigger than a cat? Am I a carnivore? Do I live in water?

You may need to model the question-and-answer process for your class. Have a student attach one card to your back, then turn your back to the students to show them what is on the card. Ask questions of the students until you figure out what is on the card. This game provides an excellent opportunity to practice deductive thinking and classification skills.

When the students have figured out which animal or plant they are, they can each remove their card and clip it to the front of their shirt. They can continue to answer questions until all of them have figured out who they are.

Next, have them sit in a circle with their cards on the floor in front of them, so everyone can see who is what kind of animal or plant.

To begin creating the web of life, take the ball of string. Holding onto the end of the string, toss the ball to a student who is portraying a plant or animal that is your predator or prey. Remember that plants derive nourishment from soil, which includes the remains of decayed animals.

Students then hold onto the piece of string and continue tossing the ball to one another. Students may catch the ball of string more than once: some animals eat many types of prey, and some are prey for many types of predators. Eventually a web is created with the trails of string from prey to predator.

When every student is holding onto a piece of the string, start a discussion about disruptions to the web of life: the introduction of pesticides into the environment, the elimination of a keystone species, or the loss of habitat. As part of the discussion, initiate a dramatic enactment of the effect of pesticides on the environment. For example, ask, "What happens when someone sprays for mosquitoes?" In response, the student portraying the mosquito falls over and plays dead. Then every student who is connected by the web directly to the mosquito falls over and plays dead. Then every student who is connected to a student who is connected to the mosquito falls over and plays dead. In this way, students see the ripple effect of environmental change and biological magnification.

Assessment: All manner of higher-level questions could be used on the next science test to assess their understanding of the food web.

Follow-Up Activities: To create a skit based on this web please see "The Song of Life" on page 87. For a more in-depth drama, students could each write a biography of one creature based on the lesson plans for "A Night on the Hunt." (See page 144.) These individual animal biographies could be strung together to create a longer novel about the food web.

Where Have All the Predators Gone?

❖**Grade Levels**: 3–12   **Time estimate**: Two 50-minute periods

❖**Science skills**: Communication; Prediction; Inference; Identify Variables; Formulate Hypotheses; Reorder, Analyze, and Draw Conclusions; Design Investigations

❖**Objectives:** Students will learn library and Internet research skills in their efforts to study predators.

National Standards

Science Standards

NAS 1 Science as Inquiry: Abilities necessary to do scientific inquiry; Understandings about scientific inquiry.

NAS 3 Life Science: Structure and function in living systems; Reproduction and heredity; Regulation and behavior; Populations and ecosystems; Diversity and adaptations of organisms.

NAS 6 Science in Personal and Social Perspectives: Personal health; Populations, resources, and environments; Natural hazards; Risks and benefits; Science and technology in society.

NAS 7 History and Nature of Science: Science as a human endeavor; Nature of science; History of science.

Language Arts Standards

NCTE 4 Students adjust their use of spoken, written, and visual language (e.g., conventions, style, vocabulary) to communicate effectively with a variety of audiences and for different purposes.

NCTE 5 Students employ a wide range of strategies as they write and use different writing process elements appropriately to communicate with different audiences for a variety of purposes.

NCTE 7 Students conduct research on issues and interests by generating ideas and questions, and by posing problems. They gather, evaluate, and synthesize data from a variety of sources (e.g., print and nonprint texts, artifacts, people) to communicate their discoveries in ways that suit their purpose and audience.

NCTE 8 Students use a variety of technological and information resources (e.g., libraries, databases, computer networks, video) to gather and synthesize information and to create and communicate knowledge.

NCTE 11 Students participate as knowledgeable, reflective, creative, and critical members of a variety of literacy communities.

NCTE 12 Students use spoken, written, and visual language to accomplish their own purposes (e.g., for learning, enjoyment, persuasion, and the exchange of information).

❖**Materials**: Pen or pencil and paper; quality topographical maps of your region; library and Internet research materials; remote video recorders, and associated equipment

Instructional Procedures

Introduction: Begin with a discussion of the role of predators in the food web and how plants and animals are adversely affected by the elimination of any species.

Discuss extinction and the effect of loss of predators on an ecosystem or food web. Begin by discussing how changes in habitat can cause large predators to become extinct. What happens then? Ask leading questions to help students deduce that without predators, the population of the prey expands until the ecosystem's food web cannot sustain it. Then the prey starves, and its population crashes. What happens then?

Activity: As a class, brainstorm a list of predators that currently live or once lived in your area. Ask students to set aside extinct animals for the duration of this activity.

You may at this point explain the difference between extinction and extirpated. Extinction means all of the animals of that type are dead. Extirpated means an animal has been eliminated from a region because of the elimination of its habitat. These animals are not extinct, because they still live in other areas.

With these concepts in mind, split the list into two categories: predators that still live in your region and those that are alive elsewhere but no longer live in your area, having been extirpated. Add more predators to each list as students mention them.

Ask students to choose one predator from either list and research its needs; the reasons it became extirpated or factors that threaten it with extirpation; food sources; habitat; range; reproduction rate; and current population. Obviously, this requires more in-depth research than some of the other projects in this book. Fortunately, many of these large animals are the subjects of extensive research; that makes digging up facts about them easier than it may at first appear. Consult with your school librarian to find sources of information. Clearly, the Internet has made this kind of research much easier. While researching this book and another book I wrote on prairie wildlife, I was able to trade e-mail with wildlife biologists, read current research papers online, view range maps, and find live video-camera feeds from nest boxes.

Challenge students to design a theoretical study that answers one of two questions: If the predator still lives in your region, what can be done to protect and promote its habitat to ensure a thriving population? If the predator is extirpated, what would be required to reintroduce the species to the area in viable numbers? Obviously, this is a theoretical study, because without millions of dollars and years of hard work, this kind of project would be next to impossible.

But It Is Being Done by Students!

At this point you may want to highlight a success story, like the reintroduction of wolves to Yellowstone National Park or the successful breeding of peregrine falcons in many major cities. In Illinois, the bald eagle population has gone from almost 0 (30 years ago) to more than 2,000 eagles sighted in the winter of 1997. The population rose from 1,733 eagles in the 1996 winter count to 2,459 in the 1997 winter count. Since this book was first published their numbers continue to climb. In 2008 there were 4,292 eagles counted! Contact your state's department of natural resources for a local success story.

Student reports should include three layers of maps color-coded to highlight current conditions as well as both short-term and long-term habitat restorations. These maps should answer the following questions: What kind of range is needed? What types of planting or streamside restoration needs to be done? If the restoration efforts are effective, what will the place look like in 10 years? In 30 years?

Have students develop graphs that reflect their species' population growth, taking into account both the age of maturity and reproduction rates. Ask students to draw two lines on each graph, one that reflects ideal conditions (based on the simple math of multiplying birthrates by the number of years), and one that reflects death rates, perdition, life spans, and other problems that are difficult to forecast.

Students could draft legislation to allocate funding and the use of public lands, or they could develop a community outreach program, including a series of lectures or PowerPoint presentations or YouTube videos or 30-second and 1-minute public service announcements that they write and produce in order to garner the support of local landowners for reintroducing extirpated species or for protecting species in danger of extinction.

Working in small groups, students could perform the initial research, and then each student could choose one aspect of the project for his or her report using a jigsaw model of cooperative learning. One student could focus on the maps, another could tackle the population charts, a third student could draft legislation and work up a budget, and a fourth could create a plan for the public awareness program.

Admittedly, this sounds like a huge challenge. But students are doing it.

Student groups in Washington State are working to restore salmon to western rivers, while their peers are helping to remove dams in Maine to restore eastern salmon runs. Students are involved in supporting the reintroduction of wolves to Yellowstone, in monitoring the tributaries of the Illinois River to help manage the reintroduction of river otters, and in raising public awareness about the manatee in Florida.

These research projects could lead to positive public support and monumental results.

Again, though they sound like expansive projects (because they are), they are not impossible tasks. When I was a high school student I wrote a letter to our governor outlining detailed plans for a summer program to hire high school students to plant trees along highways. Not only did I receive a personal reply, but he forwarded my plans to the state highway commissioner and state forestry director, who both replied to my letter, but more importantly, they initiated a roadside tree-planting program similar to my suggestions. This has been a lifelong passion of mine. I have over the past few years worked with our state's Urban Forestry director and received $10,000 to plan and organize Arbor Day events at elementary schools around the area and planted more than 10,000 trees. I also wrote a public service announcement, complete with a sing-a-long song that received airplay on several radio stations to promote tree planting.

Assessment: The research projects could be evaluated on both the quality and quantity of data, as well as the feasibility of their proposal. Student projects could be given a grade on both their individual responsibility and the overall group project. But the real success comes when songs of wolves or owls are heard at night!

Follow-Up Activities: Of all of the activities in this book, this is the one that can spark a lifelong passion for wildlife conservation and wilderness restoration

Where Does Our Food Come From? The Story of Nachos

❖**Grade Levels**: K–12   **Time estimate**: 50 minutes

❖**Science skills**: Observation; Classification; Inference

❖**Objectives**: Students will research food production and transportation to demonstrate an understanding of where their favorite foods come from.

National Standards

Science Standards

NAS 1 Science as Inquiry: Abilities necessary to do scientific inquiry; Understandings about scientific inquiry.

NAS 3 Life Science: Structure and function in living systems; Reproduction and heredity; Regulation and behavior; Populations and ecosystems; Diversity and adaptations of organisms.

NAS 6 Science in Personal and Social Perspectives: Personal health; Populations, resources, and environments; Natural hazards; Risks and benefits; Science and technology in society.

Language Arts Standards

NCTE 5 Students employ a wide range of strategies as they write and use different writing process elements appropriately to communicate with different audiences for a variety of purposes.

NCTE 7 Students conduct research on issues and interests by generating ideas and questions, and by posing problems. They gather, evaluate, and synthesize data from a variety of sources (e.g., print and nonprint texts, artifacts, people) to communicate their discoveries in ways that suit their purpose and audience.

NCTE 8 Students use a variety of technological and information resources (e.g., libraries, databases, computer networks, video) to gather and synthesize information and to create and communicate knowledge.

❖**Materials**: Food labels; pen or pencil and paper; access to library or online research materials

Instructional Procedures

Introduction: Most folks don't know where their food comes from. If you are what you eat, but you don't know where or how your food was grown, how can you expect to know what you are made of? How can you understand your role in the web of life?

Challenge students to read food labels. Are the foods they eat locally grown? (Produce is generally marked with an identifying label, or students can ask the produce manager at their grocery store where the produce comes from.) Many canned foods are labeled with "Product of." Have students list all the countries that contribute to their diet for one day.

Activity: Have students list their favorite foods, then choose one food from the list and list all of the ingredients in that food.

Ask students to choose the major ingredient of their favorite food. Using a diagrammatic flow chart, have students trace that ingredient to the store, the warehouse, the processing plant, the grain elevator, the truck, the field, and all the way back to the soil. The more boxes they have in their flow chart the easier it will be to write their stories.

Using the chart as a guide, students write a story that begins at their dinner plate and moves back through time, or starts with the seed and ends on their dinner plate. The figure on page 00 shows the progression for this activity.

Assessment: Their lists of favorite foods, the ingredients, and their flow charts can be collected and graded based on the details they provide. The story can be evaluated fro both science content and the mechanics of writing.

Follow-Up Activities: If students can scan their flow charts for illustrations, then type and edit their stories so these can be collated into a miniature picture book.

With an emphasis on locally grown products, students could put together a potluck lunch with each student providing a dish made from ingredients grown within 50 miles of their homes.

The Story of Nachos

My Favorite Food: Nachos

Ingredients

- chips: corn, salt, lime, water, oil
- tomatoes
- black olives: olives, salt, water
- green onions
- cheese: milk, rennet, salt, annato coloring

Outline for the Corn Chips section of "The Story of Nachos"

- planting
- maturing corn, maize, mythic image of young girl with blond hair, old woman bearing fruit
- tractor driving through field and picker sorting corn from silage
- truck to elevator
- barge to mill
- factory, salt, water, lime, oil in vat, fried, dried, and packaged
- truck to warehouse to store
- shopping, hungry, arrange on plate with other ingredients, microwave
- Eat!

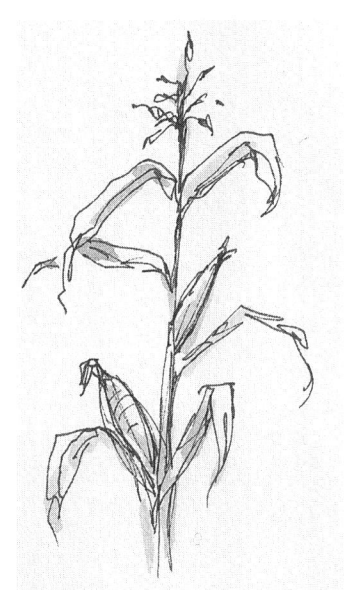

Paragraph to Start my Story about Corn Chips for "The Story of Nachos"

On a warm spring morning in Illinois, a farmer drove his tractor across a field. As he drove, his planter pierced the earth and inserted seeds of corn. The rains fell. The seed swelled with moisture. Inside the seed coat hormones were released and the tiny germ of a plant began to grow. The roots, with bits of iron, like magnets, grew toward the center of the earth. The leaves, sensing sunlight, pushed up from the ground.

From *Learning from the Land: Teaching Ecology through Stories and Activities,* Second Edition by Brian "Fox" Ellis. Santa Barbara, CA: Libraries Unlimited. Copyright © 2012.

Dragonflies and Water Quality

❖**Grade Levels**: 3–12   **Time estimate**: 50 minutes

❖**Science skills**: Communication; Prediction; Inference; Identify Variables; Formulate Hypotheses; Reorder, Analyze, and Draw Conclusions; Design Investigations

❖**Objectives**: Students will learn basic insect identification.

They will use these skills to assess water quality.

National Standards

Science Standards

NAS 1 Science as Inquiry: Abilities necessary to do scientific inquiry; Understandings about scientific inquiry.

NAS 3 Life Science: Structure and function in living systems; Reproduction and heredity; Regulation and behavior; Populations and ecosystems; Diversity and adaptations of organisms.

NAS 6 Science in Personal and Social Perspectives: Personal health; Populations, resources, and environments; Natural hazards; Risks and benefits; Science and technology in society.

Language Arts Standards

NCTE 4 Students adjust their use of spoken, written, and visual language (e.g., conventions, style, vocabulary) to communicate effectively with a variety of audiences and for different purposes.

NCTE 7 Students conduct research on issues and interests by generating ideas and questions, and by posing problems. They gather, evaluate, and synthesize data from a variety of sources (e.g., print and nonprint texts, artifacts, people) to communicate their discoveries in ways that suit their purpose and audience.

NCTE 8 Students use a variety of technological and information resources (e.g., libraries, databases, computer networks, video) to gather and synthesize information and to create and communicate knowledge.

❖**Materials**: a net (a kick-net or d-shaped net works best), a bucket to collect the specimens, an ice cube tray to help you sort the specimens, a magnifying glass or field microscope useful for identification, a field guide or poster of the bugs you are looking for (see page 77); if you are wading you might want large waterproof boots, and to report your findings, you will also need access to the Internet when you get home or return to the classroom.

Instructional Procedures

Introduction: Dragonflies really are my favorite insect, including the green darner dragonfly (*Anax junius*) that is illustrated in the children's picture book version, *The Web at Dragonfly Pond*, (Dawn Publications, 2005). Dragonflies are the hawks of the insect world. They can fly 60 miles per hour, or hover in the same spot. About 200 million years ago, there was a supersized dragonfly with a wingspan of nearly two feet across, about the size of a hawk today.

The dragonfly is also an aquatic indicator species. They will tell you if a creek is clean or polluted. Plan a class field trip to visit your local creek or pond and look for bugs. If you find dragonfly nymphs and stone fly nymphs, then it is clean. If you find no bugs in the water, it could be polluted. The kinds of insects you find tell you about the quality of the water. Your students can be science detectives and help monitor your local watershed.

To put it quit simply: the kinds of bugs you find in your local creek or pond indicate how clean the water is or isn't. In this simple exercise you can learn to identify insects that live in water. These insects will help you monitor your local stream or lake. From the types of insects you find you can make conclusions about water quality and then you can share this information with your state department of natural resources or the Environmental Protection Agency. You can make a difference in the health of the planet!

Activity: Begin with a discussion of *Aquatic Macro-Invertebrate Indicator Species.*

Say what? Take these words apart and look at the pieces. The basic concept and terminology go something like this: Aquatic means water. Macro means large . . . well, larger than microscopic. You can see them with your naked eye, but a hand lens is helpful in identification. *In*-vertebrate means no back bone. Indicator, what do they indicate? Certain species will tolerate more or less pollution. So, if the creek is very clean it will have dragonfly larvae, stone fly, damselfly, caddis fly, and mayfly larvae. If it is a little polluted it will have fewer of these and more midges and worms. If there are no bugs, *get out of the water, take a shower, and call the EPA!!!* Just kidding. Well, not really. If the creek is polluted it should be reported to your local environmental quality office.

Generally speaking, the fewer kinds of insects your students find, the more polluted the water is. The more different species you find the cleaner it is. Biodiversity is a good thing!

Because any time you are in or near water there is the danger of drowning, it is important that you have parents' permission. A good ratio of adults to children should be with you to help supervise this activity.

Go Catch Some Bugs! Collecting insects in a stream can actually degrade the stream if it is not done correctly. You do not need to turn over every rock or kick through every clump of weeds or brush. The goal is *not* to see how many things you can catch. The goal is to catch a representational sampling, just one or a few of each kind.

Turn over a rock or two in the shallows and another in a deeper area. Kick through a clump of aquatic plants with the net held downstream so whatever swims away swims into the net. Challenge students to look under roots, near the shore, in the riffles, and in the deeper pools. Collect whatever they catch into their bucket.

Some folks think it is wrong to kill any creature, so you should be careful handling the insects and turn them loose alive. Most scientists say there are so many of these creatures that if a few die in your efforts to protect an ecosystem, you shouldn't worry. This is an ethical choice you must make. I try to turn most of them loose alive.

Using the ice cube tray to help sort things out, and the field guide, poster, or the attached sheet, see how many of the insects students can find. Encourage them to make a list:

How many species can you find that are near the top of the list and indicate clean water?

_____ List a few species: _____, _____

_____

_____, _____, _____.

How many species can you find in the middle of the list indicating not-so-clean water?

_____ List a few species: _____, _____

_____

_____, _____, _____.

If you only find species at the bottom of the list, this indicates polluted water. How many?

_____ List a few species: _____, _____

_____

_____, _____, _____.

Write a Report and Contribute Your Data to an Ongoing Study of Water Quality!

Compile your information into a report on water quality that includes:

1. the date and time you completed your study

2. the name of the body of water and, if a creek or stream, what is the watershed

3. what kinds of life forms you found and in what numbers

4. your conclusion about the evidence you found

5. the names of the people involved in the study

6. an address or e-mail, contact information

Send this vital information to the state or local environmental protection agency. In Illinois, where I live, we have a statewide eco-watch program where students and teachers help scientists monitor stream quality. Many state, provisional, and regional governments have similar organizations. The U.S. Environmental Protection Agency has a Web page that lists many of the state programs so you can upload your data: http://www.epa.gov/bioindicators/html/invertebrate.html.

Assessment: Each team should turn in a report on their section of the stream.

Follow-Up Activities: For more information about using insects and other living creatures at indicator species, please visit these websites:

1. Visit your state or provincial Department of Natural Resources to find a listing of who is doing bio-indicator programs and where to report your data.

2. For more general information about bio-indicators, visit the EPA: http://www.epa.gov/bioiweb1/html/about.html

3. For a step-by-step evaluation of the process, visit this site: http://www.epa.gov/bioiweb1/html/key.html

4. Here is a detailed lesson plan written by high school teacher JoAnne Bartsch: http://www.accessexcellence.org/AE/AEC/AEF/1994/bartsch_benthic.html.

5. The Rivers Project Web page includes a wealth of information about monitoring and cleaning up your local streams. Admittedly, these folks have been a big inspiration to me: http://www.siue.edu/OSME/river/index.html.

A Handy Field Guide to Macroinvertebrates

(You may want to laminate this sheet before you take it to the creek!)

GROUP 1 – These organisms are considered intolerant of pollution

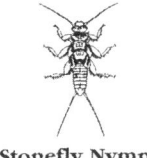
Stonefly Nymph

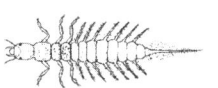
Alderfly Larva

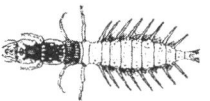
Dobsonfly Larva

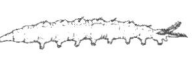
Snipefly Larva

GROUP 2 – These organisms are considered moderately intolerant of pollution

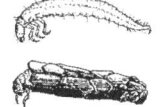
Caddisfly Larva

Mayfly Nymph

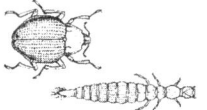
**Adult Larva
Rifle Beetle**

**Water Penny
Larvae**

Damselfly Nymph

Dragonfly Nymph

Cranefly Larva

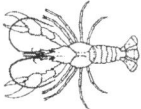
Crayfish

Clam/Mussel

GROUP 3 – These organisms are considered fairly tolerant of pollution

Black Fly Larva

Scud

**Right-Hand/
Other Snail**

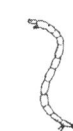
Midge Larva

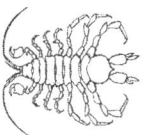
Sowbug

GROUP 4 – These organisms are considered very tolerant of pollution

Aquatic Worm

Leech

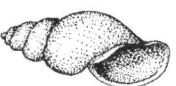
**Pouch/Left-Hand
Snail**

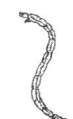
**Bloodworm
Midge Larva**

© The Rivers Project www.siue.edu/OSME/river/index.html

Take Your Students Fishing!

❖**Grade Levels**: Pre-K through Elder! **Time estimate**: A long summer day!

❖**Science skills**: Appreciation; Observation; Metric Measurement; Classification; Identify Variables; Formulate Hypotheses; Reorder, Analyze, and Draw Conclusions; Design Investigations

❖**Objectives**: Students will demonstrate a love for nature through this immersive experience.

> Because of the constant problem solving and ongoing analysis of habitat, bait, and the fish's response, students will demonstrate authentic scientific inquiry.

National Standards

Science Standards

NAS 1 Science as Inquiry: Abilities necessary to do scientific inquiry; Understandings about scientific inquiry.

NAS 3 Life Science: Structure and function in living systems; Reproduction and heredity; Regulation and behavior; Populations and ecosystems; Diversity and adaptations of organisms.

NAS 6 Science in Personal and Social Perspectives: Personal health; Populations, resources, and environments; Natural hazards; Risks and benefits; Science and technology in society.

Language Arts Standards

NCTE 4 Students adjust their use of spoken, written, and visual language (e.g., conventions, style, vocabulary) to communicate effectively with a variety of audiences and for different purposes.

NCTE 12 Students use spoken, written, and visual language to accomplish their own purposes (e.g., for learning, enjoyment, persuasion, and the exchange of information).

❖**Materials**: Hooks, line, poles, bait, tackle, scale, pen or pencil, and graph paper

Instructional Procedures

Introduction: There is no better way to involve students in the direct study of an aquatic ecosystem than to put them in the middle of the food web! At most schools you will find at least a few parents who are avid anglers and will enthusiastically share their passion for fishing. Arrange a special weekend outing with these parents. This is best done in small groups of about three students per adult.

Ask the anglers to either loan students fishing gear or help them make cane fishing poles. Also, many states have a special branch of the department of natural resources or fish and game management that will be eager to help you take your students fishing.

Activity: Catching fish is an excellent learning opportunity and encourages students' appreciation of the natural world. To build students' science skills and knowledge, ask students to try some of the following activities:

- Design studies to test the efficacy of various types of bait.

- Identify the various types of fish caught and create a chart that shows what types of fish each student caught.

- Weigh the fish and create a chart that shows the combined weight of all the fish caught.

- Create a graph that compares the number of fish caught to their total weight in kilograms.

- Dissect a few of the fish. If students caught various types of fish, dissect one fish of each type and compare their characteristics.

- Discuss the catch-and-release ethic mentioned at the end of "The Web," and encourage students to think about why it is useful or is the right thing to do.

- Have a fish fry!

Assessment: How many fish did you catch? Did you have fun trying? For the past several summers I have hosted a day of fishing for a local summer camp and though we have caught relatively few fish, almost every child says it was their favorite day at the end-of-the-week evaluations!

Follow-Up Activities: Here are several websites with tips for fishing with kids:

http://www.bwca.cc/activities/fishing/kidsfishing.html

http://www.in.gov/dnr/fishwild/3600.htm

http://www.takemefishing.org/fishing/family/little-lunkers-learning-center

Bibliography for Further Research

Arnosky, Jim. *Crinkleroot's 25 Fish Every Child Should Know*. Bradbury Press, 1993. 13: 978-0027058444. This delightful little field guide is narrated by a fictional character we would all like to meet.

Ellis, Brian "Fox." *The Web at Dragonfly Pond*. Dawn Publications, 2007. 1-58469-079-8. There was great difficulty in adapting this story from a 4,333-word epic tale, as you will find it in this book, to a 1,000-word children's picture book format. The trick was to think of it like maple syrup, boiling it down to concentrate the sweetness without losing any nutrients. Compare stories for a fun author study to contrast versions with an emphasis on writing for an audience.

George, Jean Craighead. *Acorn Pancakes, Dandelion Salad and 38 Other Wild Recipes*. Harper Collins, 1995. 13: 978-0060215491. It is highly recommended that teachers become expert at field identification and stick to plants they can definitely identify, but once you know some of the basics, and this book will help, there are some easy recipes sure to please even temperamental palates.

McDermott, Gerald. *The Stonecutter*. Viking, 1979. 13: 978-0152004002. This ancient folktale was my inspiration for "The Web." Note that both stories follow the same frame or overarching pattern. A fun follow-up lesson plan might be to read the children's book version of *The Web* and then read this book and discuss how the author was inspired by one to write the other.

The Song of Life Sings in Us All: Creating an Animal Action Poem and Skit that Celebrates the Web of Life

A scientist in his laboratory is not a mere technician: he is also a child confronting natural phenomena that impress him as though they were fairy tales.

—Marie Curie

Comments to the Teacher

RECENTLY, WHILE I WAS WORKING with a group of children aged 6 to 12 at a weekend conference in Maine, we came up with a skit to act out the local food web. Admittedly, this is an old idea, but we added a new twist: rather than being assigned roles, the kids themselves decided which animals and plants to include in the web, wrote their own parts, and then decided how best to act them out. This process required minimal coaching from me, and allowed the students a chance to thoroughly ingest the ideas and to teach each other what they knew about local ecology. This opens a whole new world of student-generated storytelling, where the process of creating story together becomes inquiry for student-led learning. Allowing your students to write the script and tell their story not only deepens their understanding of content but empowers them to take more responsibility for their education.

If one of the core principals of this book is to encourage you to tell your own story and empower your students to find their voice, this may be the lesson plan that nudges you over that line, helping you and your students to celebrate the story of your local ecosystem.

The content can be adapted to a wide range of topics. In this particular lesson you could easily adapt the play to explore a specific ecosystem—prairie, forest, wetland, desert—by limiting the choice of plants and animals to those that are native to these habitats. In this way it becomes a fun and powerful review for a study of ecosystems.

I include the story the children in Maine wrote as an example, but this is one story I will highly discourage you from reading aloud or sharing with your class as it might overly influence their creative process. I share it with you only as an example of the kind of story your students can create with the right kind of open-ended encouragement. The column on the left is the narration the students created and the column on the right in italics is a description of their pantomimed actions.

And now the story . . .

The Song of Life Sings in Us All

Introduction: (*Delivered in a conversational tone, giving the audience a chance to answer the questions out loud.*) Think about your teeth or bones. What is the dominant mineral? (*Calcium*) Where does calcium come from in your diet? (*Milk, leafy greens*) Where does milk come from? (*Cows*) What do cows eat? (*Grass, corn, and hay*) Where do they get their minerals? (*Soil and bedrock*) So inside your teeth you have sunshine, rain, soil, ancient seashells from the limestone bedrock, leafy greens, and a cow.

If you think about it, you are a lucky pile of mud! You are what you eat! You are a *lucky* pile of mud because you get to listen to poems and stories. You get to play basketball and sing out loud! Most mud just lies there, like some people you might know! As you watch this performance and listen to this poem, imagine yourself a lucky pile of mud and know that the earth's song sings inside of you!

Narration	*Action*
The Earth sings in the Soil.	*Narrator invites the audience to imagine that they are* **Soil**.
Under the soil the seeds hum softly. **Lily** and **Iris**, **Daisy** and **Aster** soak in the song of the soil, and slowly sprout, pushing up toward the song of the sun. Absorbing soil, water, and sunlight, the flowers sing!	***Lily, Iris**, and Madison sprout as flowers.*
Horses gallop through the field of flowers, nibbling the sweet nectar-filled flowers, so the song of the flower fuels the song of the horses.	*Sarah and Rachel gallop onstage as* **Horses**, *whinny, and then nibble on the flowers.*
The **cougar** stalks the song of the horses, then pounces and roars, feasting on the horse. But, eventually, he does what all of us will do (hopefully you will live a good, long life): he grows old, passes on to the other world and passes his song back to the Earth.	*Ben stalks on stage as a* **Cougar**, *gently pounces on Sara, roars and pretends to devour her. Rachel gallops away offstage. Ben then dies of old age.* *All animals quietly leave the stage when their part is over.*

Narration	**Action**
The song of the mountain lion fades to soil. **Worms** crawl in. **Worms** crawl out. **Worms** do what worms do, and the song of soil is now a richer song.	*Cam and Syrus squiggle onstage as **Worms**. They crawl over the cougar and then make loud farting noises before crawling off.*
The **Poison Ivy** grows richer and greener in this fertile soil and sings an itchy song.	*Ian begins to sprout as **Poison Ivy,** wrapping a tendril around imaginary branches.*
Deer are not allergic to poison ivy. It is an important food in the late autumn. So now you hear the song of the deer.	*Syrus meanders onstage as a **Deer** and then eats the ivy with loud chomping sounds.*
A **Horsefly** buzzes around the deer, lands on his back, and takes a bite. The blood of the deer feeds the song of the fly.	*Rachel the **Horsefly** buzzes around, lands on the deer, and bites him. She flies away and lands on a branch.*
When the fly lands on a twig to rest for the night, a **Lightning Bug** lands on her, devours her, and the song of the fly adds light to the bug!	*Max comes onstage with a flashlight blinking. As a predatory **Lightning Bug** he eats the horsefly. (This is stretching it, but some lightning bugs are predatory and Max was a cute four-year-old who wanted to be a lightning bug.)*
A **Frog** hops over to the edge of the meadow, singing a frog song. She whips out her long tongue and swallows the lightning bug, singing a frog song.	*Sarah, the **Frog**, hops out, whips out her tongue and swallows the lightening bug. She ribbits.*
A long black **Snake** quietly slithers up on the unsuspecting frog, lurches forward, and the croak of the frog adds to the hiss of the snake.	*The **Snake**, Cam, slithers onstage, opens his mouth wide, and swallows the frog. He hisses.*
A **Housecat** stalks the snake, pounces upon it, and carries it home to her master, meowing and purring the song of a successful hunt.	*The **Cat**, Megan, crawls out quietly, pounces, meows, and carries the snake to her master.*
Her master, a **Human**, sings a song of praise. He takes the snake inside, slices it up and fries it for supper. He tosses part of it to his trusty dog.	*A **Human** named Ian gently strokes his favorite cat and says, "Thank you, kitty. Good kitty. I was wondering what I was going to have for dinner."*
The **Dog** gulps down the table scraps. He barks and howls a song of gratitude.	*Ian tosses a scrap to the **Dog**, Cody, who barks and howls before walking over to the oak tree and lifting his leg.*

Narration	*Action*
He wanders out into the forest and waters a tree. You might think this is gross, but the **Oak Tree** drinks in this fertile water. The minerals help the tree to grow, whispering quiet songs in an evening breeze.	*The **Oak Tree**, Whitney, makes a slurping sound and raises her branches, pretending to drop acorns.*
As the tree drops its mast crop of acorns a large **Black Bear** eats these acorns to put on weight for the coming winter. The song of the tree becomes the roar of the bear!	*Jessica, a **Black Bear**, ambles on-stage, eats acorns, and then growls a loud growl.*
And the song of the Earth sings in you! Inside of you are bits of bear and tree, dog and snake, frog and fly, deer, ivy, and worm. Cougar, horses, and fields of flowers all sing their song inside of you. Celebrate the songs of the Earth, sing the joy that flows through your veins.	*As each character's name is mentioned, he or she stands up and takes a bow. At the end, all bow together.*

Follow-Up Ideas for "The Song of Life"

Student-Led Drama

❖**Grade Levels**: K–5   **Time estimate**: 45 minutes

❖**Objectives**: Students will demonstrate a knowledge of the food web in the creation of a skit that includes producers and consumers.

Students will build public speaking and acting skills.

National Standards

Science Standards

NAS 3 Life Science: Structure and function in living systems; Reproduction and heredity; Regulation and behavior; Populations and ecosystems; Diversity and adaptations of organisms.

NAS 6 Science in Personal and Social Perspectives: Personal health; Populations, resources, and environments; Natural hazards; Risks and benefits; Science and technology in society

Language Arts Standards

NCTE 4 Students adjust their use of spoken, written, and visual language (e.g., conventions, style, vocabulary) to communicate effectively with a variety of audiences and for different purposes.

NCTE 5 Students employ a wide range of strategies as they write and use different writing process elements appropriately to communicate with different audiences for a variety of purposes.

NCTE 11 Students participate as knowledgeable, reflective, creative, and critical members of a variety of literacy communities.

NCTE 12 Students use spoken, written, and visual language to accomplish their own purposes (e.g., for learning, enjoyment, persuasion, and the exchange of information).

❖**Materials**: Paper and pencil

Instructional Procedures

Introduction: This lesson stands well on its own, but is also a great extension of "Adventures in the Web of Life," page 63.

Activity: Here is a synopsis of the lesson:

1. First we made a list of students' names and their favorite plants or animals.

2. Next we discussed the food web, consumers, producers, and decomposers.

3. Then we outlined the sequence of events, discussing the order of actors in the skit.

4. I took notes on the entire process and read the notes back to the students for discussion and rewriting.

5. We rehearsed twice and then we performed.

6. Altogether, the entire process took little more than an hour. We spent about 45 minutes in our initial discussion and first run-through, and then about 10 minutes in each of two rehearsals before presenting the skit to an audience.

The following lesson plan looks at each of these steps in more detail.

Choosing Characters

Sitting in a circle with the teacher taking notes, the first thing we did was pass a *talking stick*. As the stick was passed to them, the children said their names and the name of their favorite plant or animal. I encouraged them to choose plants and animals from the local ecosystem. I challenged them to be creative and think of unusual creatures, insects, favorite flowers, birds, reptiles, and fish, not just cuddly mammals or the classic charismatic megafauna, lions and tigers and bears, oh my! Because we were focusing on local ecology, I also made one restriction: the plant or animal had to be one that they had actually seen around their hometown or farm. It was fine if two or three students wanted to be the same animal; that just meant that we had a herd or flock of them. As the students took their turns, I wrote down their names and their favorite plant or animal. This was our initial meeting as a group and a good way to begin to get to know each other.

After we had compiled our initial list, we discussed consumers and producers. Who eats whom? In a Socratic style, I asked about herbivores, carnivores, and omnivores, to find out what the students knew and to allow them to teach each other. I introduced the concept of decomposers, or detritivores. Then we passed the talking stick again. Each student restated his or her favorite plant or animal and, if the choice was an animal, stated whether it was herbivore, omnivore, carnivore, or detritivore. All the while I was taking notes.

Once we had our revised list of characters, we discussed the order of events in the skit, based on the question of who eats whom. Which animals or plants are on-stage first? Who might eat them? Who might eat *them*? As we began mapping out the sequence, we noticed gaps: for example, what is the trophic chain that links a tree to a cougar? (Yes, there are cougars in Maine!) Without much encouragement needed, students suggested new animals, plants, or insects to fill in such gaps. As we started walking through the skit, the students felt as if they were in charge and were very eager to make suggestions and refine the presentation.

The teacher's role throughout this process is to ask thoughtful questions, without leading the students by the nose, so they create and have ownership of the characters, the action, and the entire story line.

Creating the Narration

The next step was to create the narration. I had taken notes on the students' discussion throughout the process. In this way they helped me to write the script. I simply did a minimal amount of editing of their self-directed actions. I introduced the "Song of Life" refrain, but once we got this idea rolling they picked it up and began inserting it on their own. My goal was to use their words as much as possible and let them solve the problem of moving the play forward.

It was sometimes hard for me to avoid jumping in with directions, but I tried to lead the discussion with Socratic questions and let them do the thinking. For example, when one of the children said, "The cougar pounced down on the horses," I wrote this down and then asked: "When you say pounced down, down from where? Please show me what that looks like! As you are acting this out, please talk. Tell me what you are doing so I can write it down." As I read back to them what they created, I encouraged them to think of refinements, to make edits and suggestions. I also encouraged them to sing, growl, whistle, howl, and add a lot of sound effects to their pantomime. Of course, the sound of the worm defecating was a big hit!

The students' ownership of the script meant that they did not have to memorize or study their parts because they had created them. We rehearsed twice, running through the play on both Saturday and Sunday afternoon. Each time, I simply read the narration, which cued them to act. (It would be just as easy, and maybe more effective, for a strong-voiced student to read the narration.) Sunday evening we performed for the parents as the warm-up act for a folk music concert. I narrated and they acted out the poem in pantomime with hilarious sound effects.

This activity could easily be done in a school setting, with one class writing and developing the production and performing it for other classes or for a family night. At an environmental education center, different groups could write similar skits about different ecosystems: one could act out a wetland or prairie food web and another could do a forest or desert. That evening they could all be performed with variations of the song "Mother Nature Had A Prairie" (see page 93) as a segue between skits.

The narrative poem and skit that my students helped to write is presented here as an example. Although you may wish to use the introduction, do not read the script to your students; rather, let each group of students share the learning and the fun of creating and singing their own "Song of Life."

Assessment: The applause or lack thereof from the audience is all the evaluation needed!

Follow-Up Activities: Take the show on the road! This skit would be fun to perform for other classes, or for a special family night that deals with science literacy. After the class has walked through the process, they could be divided into smaller groups and each group write their own script about a different ecosystem.

Take an Old Song and Make it New

❖**Grade Levels**: K–12   **Time estimate**: 10–50 minutes

❖**Science Skills**: Communication

❖**Objectives**: Students will rewrite a traditional folksong to explicate an important science concept.

National Standards

Science Standards

NAS 3 Life Science: Structure and function in living systems; Reproduction and heredity; Regulation and behavior; Populations and ecosystems; Diversity and adaptations of organisms.

NAS 6 Science in Personal and Social Perspectives: Personal health; Populations, resources, and environments; Natural hazards; Risks and benefits; Science and technology in society.

Language Arts Standards

NCTE 12 Students use spoken, written, and visual language to accomplish their own purposes (e.g., for learning, enjoyment, persuasion, and the exchange of information).

❖**Materials**: Paper and pencil

Instructional Procedures

Introduction: Folksongs have been around for thousands of years and every generation makes them their own, rewriting and updating them to match the mood or use the slang of the current cultural milieu. Everyone loves to sing, whether or not they will admit it! Taking old folk tunes and rewriting them as environmental ballads is a fun way to reiterate core science concepts in a different vocabulary.

The idea for this song is based on a version of "Old MacDonald Had a Farm" that I heard from a puppeteer in North Carolina more than a dozen years ago. With beautiful handmade puppets she acted out the food web of a swamp, and all the while the audience was singing along with "Mother Nature Had a Swamp!" My daughters and I have made up several incarnations with several ecosystems and their representative animals. My daughters Lily and Laurel helped me to write this song while driving to a performance at the Brookfield Zoo. We changed E-I-E-I-O to Hey-ya-hey-ya-ho, so it reflected our Native ancestry. Folks love to sing folk songs; please sing-a-long!

Activity: Sing the song "Mother Nature Had a Prairie."

Challenge students to work in small groups of four to first choose a favorite folksong or popular tune and rewrite the lyrics to explore their favorite science concept.

For early elementary students they can simply rewrite "Mother Nature Had a Prairie" to explore a new ecosystem, forest, tundra, swamp, or desert.

For older students they can work in groups of four to choose a folksong they all agree on, or pick a modern rap, rock, or country tune by their favorite artist and then rewrite the lyrics to explain a science concept. Though writing by committee can be tricky, if they work well together they can bounce ideas back and forth to write a witty song. If a group is having trouble, its members can each write a verse to the song and then take turns singing the verses with everyone joining in on the chorus.

Assessment: The lyrics can be collected for a grade and their performance can be evaluated based on a standard rubric.

Follow-Up Activities: Once again I say, take this show on the road! Older students can take their songs on tour and perform for younger classes. These kinds of songs are great for a family night! Students who play an instrument could work with a music teacher and be encouraged to write original music and lyrics.

These songs can be recorded as audio or video files and burned onto a CD or posted on YouTube!

Bibliography for Further Research

Rogers, Sally, illus. Melissa Bay Mathis. *Earthsong*. Dutton Children's Books, 1998. 0-525-45673-5. This is a perfect example of taking an old song and making it new! Sally Rogers has rewritten the folksong "Over in the Meadow," creating an inspiring sing-a-long about the Earth's endangered species.

Schnetzler, Pattie, illus. Chad Wallace. *Earth Day Birthday*. Dawn. 2004. 13: 978-1584690542. Admittedly, when I started reading this book and recognized the song they used as a frame was "The Twelve Days of Christmas." I did not think it could work, but it does work, beautifully. With clever lyrics and stunning pictures they have made magic by bringing us into an intimate celebration of twelve creatures' birthdays, an Earth Day anthem that is a nice frame for students to create their own verses!

Mother Nature Had a Prairie

Rewritten by Brian "Fox" Ellis
(Sung to the tune of "Old McDonald Had a Farm")

Mother Nature had a prairie, Hey-ya-hey-ya-ho

And on that prairie she had some snakes,
  Hey-ya-hey-ya-ho

With a hiss-s-s here and a hiss-s-s there,

Here a hiss-s-s, there a hiss-s-s, everywhere a hiss-s-s,

Mother Nature had a prairie, Hey-ya-hey-ya-ho!

Mother Nature had a prairie, Hey-ya-hey-ya-ho

And on that prairie she had some coyotes,
  Hey-ya-hey-ya-ho

With a ho-o-owl here and a ho-o-owl there

Here a ho-o-owl, there a ho-o-owl, everywhere a ho-o-owl

With a hiss-s-s here and a hiss-s-s there,

Here a hiss-s-s, there a hiss-s-s, everywhere a hiss-s-s,

Mother Nature had a prairie, Hey-ya-hey-ya-ho!

Mother Nature had a prairie, Hey-ya-hey-ya-ho

And on that prairie she had some prairie chickens, Hey-ya-hey-ya-ho

With a cluck-cluck here and a cluck-cluck there,

Here a cluck, there a cluck, everywhere a cluck-cluck,

With a ho-o-owl here and a ho-o-owl there

Here a ho-o-owl, there a ho-o-owl, everywhere a ho-o-owl

With a hiss-s-s here and a hiss-s-s there,

Here a hiss-s-s there, there a hiss-s-s, everywhere a hiss-s-s,

Mother Nature had a prairie, Hey-ya-hey-ya-ho!

You get the idea? I usually let the kids choose the next animal and help me with the sound effects. They notice the pattern quickly and automatically join in. It is easy to change the word *prairie* to forest or desert or swamp or ocean and then change the animals accordingly.

From *Learning from the Land: Teaching Ecology through Stories and Activities,* Second Edition by Brian "Fox" Ellis. Santa Barbara, CA: Libraries Unlimited. Copyright © 2012.

Making a Simple Food Web

(For a more complex version of this activity please see "The Web" on page 65.)

Materials: A ball of yarn or twine

To make the idea of the food web more graphic and tangible, this activity works best after the script is written and the students know their parts. While you are rehearsing the skit with your class, sit everyone in a circle with a large ball of twine or colorful yarn. As you read the narration, ask students to take turns standing up and acting out their parts while staying in their places within the circle. Ask students to roll the ball of yarn across the circle to the person or animal that eats them when they are done acting out their parts. If rolled correctly the ball of yarn will leave a gossamer thread that shows the trophic cycle. At the end you should have a beautiful spiderweb stretched out across the floor.

Through Socratic questioning, lead a conversation about this web: What do these strings show? When you look at all of the strings together what do you see? Why do you think *food web* is a better idea than *food chain?* What would happen if we take out one character? Who would not have anything to eat? Then what happens?

Explore the Song of Life in Your Backyard!

❖**Grade Levels**: K–5   **Time estimate**: 45–60 minutes

❖**Science Skills**: Observation, Communication, Inference

❖**Objectives**: Students will learn to observe signs of wildlife and make inferences about animal behavior and the relationships between consumers and producers.

Students will demonstrate creative writing skills in the creation of stories about the food web in their backyard!

National Standards

Science Standards

NAS 1 Science as Inquiry: Abilities necessary to do scientific inquiry; Understandings about scientific inquiry.

NAS 3 Life Science: Structure and function in living systems; Reproduction and heredity; Regulation and behavior; Populations and ecosystems; Diversity and adaptations of organisms.

Language Arts Standards

NCTE 5 Students employ a wide range of strategies as they write and use different writing process elements appropriately to communicate with different audiences for a variety of purposes.

NCTE 7 Students conduct research on issues and interests by generating ideas and questions, and by posing problems. They gather, evaluate, and synthesize data from a variety of sources (e.g., print and nonprint texts, artifacts, people) to communicate their discoveries in ways that suit their purpose and audience.

❖**Materials**: Clipboards, a few sheets of unlined paper, pencils, crayons, or colored pencils

Instructional Procedures

Introduction: After a performance of "The Song of Life," discuss with students the food web in their backyard. How might they find evidence of these relationships between predator and prey? Do they think they will see a cougar eating a deer? What are some signs of life they might find?

Activity: Take your class outdoors! In your school yard or neighborhood there are lots of signs of the song of life! Lead a short hike, maybe five minutes, where you ask them to help you find signs of predator–prey relationships. Who is eating whom?

It could be as simple as: a leaf with holes in it where some insect has been munching; or earthworm castings where worms have been eating the bacteria that live in the soil; or an acorn rind where a squirrel was eating an acorn; or an ant carrying home the carcass of another insect. Sometimes you will see the telltale signs without seeing the actual devouring, like a fluff of scattered feathers, where a bird was eaten; or a peeled walnut husk where a chipmunk had been eating. Like detectives at the scene of the crime, piece together the clues. Begin to outline a story line as you lead the hike, inviting students to help flesh out the narrative.

Depending on your school grounds or neighborhood and the age level of your students, allow me to encourage you to turn them loose for 5 or 10 minutes to find their own signs of the song of life. With clear limits on boundaries, tell them that they only have 2 minutes to find a scenario, and 3 or 4 minutes to make a quick sketch. Yes, it is okay if two or three people are drawing the same scene. But remember, they might have to fill in the blanks. If they see a leaf eaten by an insect tell them to go ahead and draw the insect that might have eaten the leaf. If they see a spider's web, tell them to go ahead and draw what the spider might look like, even if she isn't home.

When the time is up, draw the class back together and have each student share pictures with a partner, describing the scene and then writing a few sentences at the bottom of the picture that might be part of a larger story.

If the weather is nice you can finish outdoors, or if not, go back to the room for the rest of the activity.

Explain that artists, as well as writers, make a rough draft or what is called a quick study. Pass out the crayons or colored pencils and ask students to take their time and reimagine the scene. Draw the key elements large and central. Add a background that might add detail to the story. If a picture is worth a thousand words, then it should be easy to write an exciting paragraph as a caption for the picture. With younger students you can ask parents to take dictation and help with spelling, as long as they put only the student's words on the page. In either case the caption can be typed, clipped, and pasted onto the picture, top or bottom.

Assessment: Teachers can collect student pictures, their rough drafts, and their finished projects, which can be evaluated for both the process of observation and inference as well as the final product based on scientific accuracy and the mechanics of language arts.

Follow-Up Activities: These pictures can then be used for a bulletin board, bound together as a book, or used as story starters as a follow-up to "A Night on the Hunt." (Please see page 144.)

Bibliography for Further Research

Arnosky, Jim. *Crinkleroot's Guide to Walking in Wild Places*. Bradbury Press, 1990. 13: 978-0689717536. Another great book in the *Crinkleroot* series that is a great companion for a walk around the school yard as well as a local park.

The Rattlesnake that Tamed a Boy: A Tall Science Tale

Never a day passes but that I do myself the honor to commune with some of nature's varied forms.

—George Washington Carver, from *George Washington Carver*

Comments to the Teacher

ALTHOUGH THIS IS A TALL tale, tell it straight, as if even the most fantastic detail was a plain, everyday thing. If you sense that your credibility is waning, play with your audience. I often use phrases like, "I know this is hard to believe, but . . ." Then I state a credible scientific fact, such as, "Snakes are territorial and have been known to travel miles to return to their turf. How many of you know the story of *The Incredible Journey*? If two dogs and a cat can travel 200 miles, don't you think a snake could travel 30?"

Use body language and vocal expression to dramatize the story, but keep it subtle. Part of the fun of a tall tale is suspending disbelief, allowing yourself to be fooled into thinking the fiction is fact. Later, students can sort the fact from the fiction. But in the telling, let it all be true!

If you start with the truth, people believe you. To help them suspend their disbelief, stretch the truth, oh, so gently. To make it work for you, consider rewriting that part of the story to reflect your own childhood experiences. Adapt the story to use one of your relatives who lived on a farm in a part of the country that has rattlesnakes. If you do not have such a relative, use my uncle as if he were yours. If this is uncomfortable, tell this story in the third person. Pick a name for the main character, and tell the story about that person. Consider using a girl's name for the main character. Just because the character is a crack shot doesn't mean it has to be a boy!

This can be a long or short story, depending on how much time you have to tell it. I have often told 45-minute versions of this story with a great deal of detail about my childhood and several tangents that covered intriguing facts about snakes. I have also told 10-minute versions of the story that followed the core events. If time is short, cut several scenes or issues, such as petting the snake, the snake in the house, the snake in the bed, or leaving the snake in the woods. But be aware that each time you eliminate a scene you lose an opportunity to inject humor and intriguing science facts, so cut carefully.

Note: To add an audible element to the story, make a rattlesnake sound whenever the word *rattlesnake* comes up in the story. To make the sound, rapidly drum your tongue against the roof of your mouth: *t-d-d-d-l*. Or shake a baby rattle or jar with several beans in it. After a few repetitions, your students will anticipate the sound and join in.

And now the story . . .

The Rattlesnake that Tamed a Boy: A Tall Science Tale

One of the ways in which science works is through contrast and comparison, discerning this from that, myth from fact, useful from destructive. As you listen to this story, compare and contrast city and country, natural and artificial. What is good and bad in each?

This is a story about a pet snake I had. As you listen, think about what you know about snakes. What do you believe about them? What myths about snakes have you believed at one time or another?

250 = how many? Seventeen hundred acres! That was my backyard!

My uncle grew soybeans and corn and hay. He also had 180 cows. Can you believe that? He had a huge dairy operation, with electric milking machines and enormous refrigeration tanks. My uncle drove the milk around in stainless steel tanker trucks. Most people only see milk in plastic jugs from the store, but that summer I got to see the whole process, from cow to supermarket.

When I was nine years old my uncle invited me to spend the summer on his farm in Tennessee. What a dream come true: a whole summer far away from the noise and smells of the city. In the city, I had a backyard that was 30 feet by 40 feet—a tiny fraction of an acre. At my uncle's farm, I would have more than 1,500 acres to play on. Here's how it worked: My uncle owned 350 acres. The neighbor to the east owned 700 acres, the neighbor to the north owned 400 acres, and the neighbor to the west owned 250 acres. Do a little math with me: 350 + 700 + 400 +

The only thing that bothered me was my fear of snakes. I love snakes now—and soon you'll see why—but before I went to the farm, I was afraid of them.

I was especially afraid of *(t-d-d-d-d-l)* rattlesnakes.

When I was out and about, I always kept my eyes and ears open for snakes. One day when I was walking along the edge of the woods, where the woods meet the pasture, *t-d-d-d-d-l*, I heard a rattlesnake. I froze. Have you ever been so afraid that you want to scream but nothing comes out? Have you ever been so afraid that you want to run but you are frozen with fear? That's how I felt.

Without moving my feet or body, I turned my head to look around. About 20 feet away, there was that rattlesnake. It was hissing and shaking its tail. It was the strangest thing I'd ever seen. Let me tell you what I saw: The snake was sandwiched between two rocks. The rock on top pinned the snake so it couldn't get away. But the snake was still alive and didn't seem to be hurt. It certainly wasn't too hurt to rattle! *T-d-d-d-l.*

Let's stop to think about this. How did the snake get into that predicament? What do you think? This is a science story; help me formulate a hypothesis. More than an educated guess, it is drawing a testable conclusion based on what you already know. What do you know about snakes? What do you know about the situation? Using this information, make a guess.

[*Give students time to discuss this with a partner. Think, Pair, Share. Then, saying something like, "Let's hear some of your hypotheses," call on a few students. In offering feedback on their hypotheses, focus more on affirming their logic and reasoning than on a correct answer.*]

Honestly, I don't know for sure how that snake got stuck, but here's my hypothesis:

Snakes are nocturnal, right? They come out at night and sleep during the day. Are you nocturnal or diurnal? Right, diurnal. You come out during the day and sleep at night. That makes you diurnal.

Snakes are also cold-blooded, right? Well, we now know they are polythermic, which means they can generate some heat while digesting, but generally, they are cold-blooded or ectothermic. They cannot make their own heat, so they are always looking for a warm place. Are you ectothermic, cold-blooded, or endothermic, warm-blooded? Warm-blooded, right. Maybe the snake was sleeping on a sunny rock. And maybe a cow walked by farther up on the hill. The cow stepped on a rock, and the rock rolled down and landed gently on the sleeping snake.

Is this like your idea? Maybe you are right and I am wrong, maybe I am right and you are wrong. We'll never know for sure.

What we do know is that snake was stuck, and so was I. What was I going to do? I couldn't just leave it there. I was afraid of snakes in general, but I felt sorry for this one. I felt like I had to help it in spite of my fear. What was I going to do? What would you do if you were in my shoes?

I thought and thought, and then I looked around me and had an idea. I went into the woods and found a long stick. Standing as far away as I could, I used the stick like a crowbar to pry the rock off the snake.

The snake was loose!!! It slithered out from underneath the rock. It seemed to be okay, but I have to say I didn't go up close, to look. The snake squirmed some, and then it started to come toward me. Did I tell you I was afraid of snakes?

"Good-bye, nice helping you," I called over my shoulder as I took off. I walked as fast as I could, but the snake followed right behind. I started running. I ran as fast as I could, but it still came after me!

After a while I started to get tired. I slowed down, but kept on walking fast. The snake came up behind me. It was then I noticed that the snake never came too close to me. It always stayed just out of reach, as if to prove that it did not want to hurt me. I looked over my shoulder at it. Why would that snake be following me like that?

I love a good story. Sometimes they give us answers to our questions. Do you know the story of the mouse and the lion with the thorn in its paw? When I saw that snake following me it made me think of that mouse and lion, how the lion promised not to hurt the mouse because he had saved its life. My imagination went wild. Wouldn't it be great to have a pet rattlesnake?

Gradually, I stopped worrying about the snake. It never came close enough to bite me. My fear slithered away as my excitement grew. I kind of hoped the snake would follow me all the way home.

And it did! When I reached the farm, my uncle was outside. I walked right up to him, with that snake following behind me, and I said, "Sir, if some wild animal followed me home and wanted to be my friend, could I keep it as a pet?"

"Sure, if that's what you want." Just then he noticed the rattlesnake. He pushed me out of the way and raised his ax to chop off its head.

I jumped into his way. "No, no, please don't hurt it. That's my pet."

My uncle shook his head. "I thought you were a little crazy, boy, but I didn't know how crazy you were. Well, I said it was okay and I'll stand by my word. But if it causes any trouble, it will have to go." My uncle always kept his word, even when he wasn't happy about it.

I must admit I was afraid at first that the snake might bite me, but it never did. If the truth be told, that snake was a good pet because it ate the rats and mice underneath the house and around the barn. My uncle liked this, a lot. If you have ever lived on a farm, you know that no matter how clean your house is, you are going to have rats and mice. They are attracted to animal feed.

I knew something about animal feed because one of my chores was to feed the chickens. The rule on the farm was, I could not eat breakfast until the animals were fed. So early every morning I would go out to feed the chickens,

and every morning some of their food would be missing—until that rattlesnake moved in. Whether the snake ate all of the mice or whether its scent and presence scared them away, I am not sure. All I know is the food stopped disappearing, and my uncle was happy about that.

I will never forget the first time I touched that snake. Have you ever touched a snake? Some people think snakes are slimy. I once thought so, too, but if you have ever touched a snake, you know they are not slimy at all.

This is how I came to touch the snake: The snake always slept in the sun on the front steps of the house. Remember, snakes are nocturnal and cold-blooded. They like to sleep in the sun.

One day, I sat on the front steps next to the snake. It did not wake up. I sat there for the longest time, working up my courage. I reached . . . but I couldn't do it. I had never touched a snake before. I reached down . . . but I couldn't do it. Finally, I screwed up my courage. I reached down and oooh! It wasn't at all what I expected; it was cool, and smooth like leather.

S-s-s-s-th-th! The snake woke up. S-s-s-sth-th. It liked the touch of my warm-blooded skin. It crawled up my hand. It wrapped around my arm and went inside my sleeve! Its head poked out of my collar behind my head and then went down the other shoulder and out the other sleeve. I was ready to scream, but then I thought, "Cool, a snakeskin collar!" It did feel great, like a back rub or shoulder massage. If you think about it, snakes are one long string of powerful muscles wrapped around a backbone. No wonder it felt good.

After that I started picking up the snake whenever I wanted to.

I will never forget the first time I brought the snake into the house. We were sitting around the kitchen table one night, eating fried chicken, mashed potatoes and gravy, collard greens, and corn on the cob, when my uncle told my aunt about the snake eating the mice in the barn.

She said, "Why, you ought to bring that snake in here. I saw a mouse in the kitchen just the other day." I didn't know she was just joking.

I had seen that mouse myself. It had a hole behind the refrigerator. It would crawl up through the wall and eat our cereal, cookies, and crackers in the cupboard. The next day, when my aunt went to the grocery store, I decided to help out. I went outside and got the snake and brought it into the kitchen.

Right away, the snake started to stick out its tongue. S-s-s-s-th-th. Do you know why snakes stick out their tongues? One reason is, snakes use their tongue to smell. How many nostrils do you have? Two, right? So does a snake. But a snake has a secret advantage. Here's how it works: A snake sticks out its forked tongue to pick up scent particles in the air. Then it puts its tongue back in its mouth. So far, so good, right? But get this: then the snake pokes the two ends of its tongue into its nostrils— from the *inside of its mouth*. In this way snakes can smell things that you and I cannot. It tastes its environment.

Snakes do have an incredible sense of smell. Right away, my snake smelled where that mouse had been. It headed straight for the refrigerator, then crawled up into the wall. You could hear the snake heading up through the wall

over toward the cupboard, following the route the mouse took.

Just then my aunt came home. She set the grocery bags on the counter. She put the milk in the refrigerator. She went to put the cookies in the *cup—A-a-a-h! [Shriek]* She almost lost her cookies, if you know what I mean.

Well, the mouse disappeared. Whether the snake ate it or the smell of the snake scared it away, I'm not sure. All I know is that we did not see any more mice in the kitchen.

After that, the snake came into the house on a regular basis. I will never forget the first time it crawled into bed with me. It was early morning, and the snake must have finished hunting for the night. I was still sleeping when it crawled into my bed and curled up around my feet. At first it tickled, and I thought I was dreaming. Then it started squeezing, and I woke up. Boy, was I scared!

I lay perfectly still. I thought if I moved, the snake would bite me and I would die. Now I know that most snakebites are not deadly. If a large snake bites a small person near the heart, watch out. But if a snake bites you on the hand or the foot, you won't die. Your hand or foot will swell up like a football, but usually you don't die. In either case you do need immediate medical attention!

Finally, I got up the courage to slowly remove the snake. It didn't bite me; it didn't seem to mind at all. After that, the snake slept by my feet pretty often. He kept my feet warm, and I warmed his cold body. It never bit me and I never bit it.

As the summer drew to a close, I began to worry about the snake. I couldn't take it back to Toledo—that's no life for a snake. Finally, I decided if it was time for me to go home, the snake should go home, *too.*

The next day I went for a long walk, and the snake went with me. I sat down in the middle of a huge meadow. The snake curled up next to me and fell asleep in the sun. I waited to be sure the snake was sound asleep, then I quietly sneaked away. When I was some distance off, I started to run. I ran all the way to the farm, thinking, "Ha, ha! I really tricked that snake! It will never find its way back now!"

But I was also sad to think I'd never see the snake again. He was my best friend from the summer. Thinking about all the fun we had, I climbed the porch steps and who do you think was there? That snake must have known a shortcut, because it got home before I did!

That night I told my uncle what I had done. He said, "You had a good idea, but you weren't thinking big enough. But you're right, you are leaving soon, and we can't keep the snake after you're gone. Let's give it another try tomorrow."

I didn't like the idea, but I had to agree that my uncle was right. The next day we packed a picnic lunch in a basket. We put the snake into the backseat of the car, and we drove off. We drove for more than an hour. Then we had a picnic lunch. After lunch, we played some games, and then we curled up as if we were going to take a nap. The snake curled up and went to sleep. We got up quietly and climbed into the car. We didn't slam the door because we didn't want to wake up the snake.

It was the saddest ride of my whole life. Has your best friend ever moved away? Have you ever had to move away from your best friend? Can you remember or imagine how that felt? That's how I felt. I cried all the way back to the farm. My uncle was a smart man; he knew that sometimes you just have to cry. When we got back to the farm, I went straight to bed and cried myself to sleep.

In the morning I felt better. I thought about the good times we had going for walks in the woods, the snake crawling into my bed, and that first time I brought the snake into the house, when it scared my aunt.

I felt a little sad, but I knew the snake was better off in the woods. Not only did I feel better, I was hungry, too. Remember, I had gone to bed without supper. But you know the rule: no one eats until the animals are fed. So I went outside to feed the chickens . . . and tripped over something on the porch. Now what do you suppose that was?

Right! The snake. Wow! I could hardly believe it! I was so happy that I whooped and hollered.

My aunt and uncle came running to see what was wrong. My uncle took one look at that snake and said, "If that snake is going to travel that far to live in this house, it can live here forever. I'll take care of it when you go back to Toledo."

Well, the snake and I had a great time our last

week together. But don't think that's the end of the story.

There's something I forgot to tell you. As you know, that was my first pet rattlesnake. When you have a pet dog you can train it to scratch on the door when it wants to go outside, to roll over, or give you a paw, or not to wet on the carpet. Well, I had never had a pet snake before, and I didn't know how to train it. So, it picked up a bad habit. I wish I had trained it better, but I didn't know how.

The snake had a bad habit of chasing cars. Do you believe me? No? Well, I don't expect you to believe that the snake would chase cars. All right, the snake wasn't *really* chasing cars. But the dogs would chase the cars and the snake would chase the dogs, so it looked like the snake was chasing the cars. It was actually quite funny if you could picture it. But the end was inevitable.

One day I came out onto the porch, and the snake wasn't there. I looked around the yard. It wasn't there, either. Then I looked around the house. It wasn't there, either. Usually, the snake stayed on the porch or in the yard, because it liked to sleep in the sun. But today it was nowhere to be found. I knew its bad habit, and I had a sinking feeling that the inevitable end had finally come.

I went out through the gate and down the dirt drive to the road. I looked down the road, but I didn't see it. Then I looked up the road, but I didn't see it there, either. Then I saw it, squashed flat in the middle of the road. It was chasing a car and it got run over and it was lying there squashed flat, just lying flat out, just lying there, *like I've been lying to you the whole time!*

Follow-Up Ideas for "The Rattlesnake that Tamed a Boy"

Discussion Topics

The obvious question is this: Where does the truth end and the tall tale begin? Parts of this story are true and some of it is patently false. My favorite irony in this story is that I am using a tall tale to dispel myths about snakes! This general discussion leads right into the first lesson.

Discerning the Facts

❖ **Grade Levels**: 2–12   **Time estimate**: 45–60 minutes

❖ **Science skills**: Observation; Communication; Inference; Formulate Hypotheses; Re-order, Analyze, and Draw Conclusions; Design Investigations

❖ **Objectives**: Students will sort fact from fiction and model civil debate in their discernment.

National Standards

Science Standards

NAS 1 Science as Inquiry: Abilities necessary to do scientific inquiry; Understandings about scientific inquiry.

NAS 3 Life Science: Structure and function in living systems; Reproduction and heredity; Regulation and behavior; Populations and ecosystems; Diversity and adaptations of organisms.

NAS 6 Science in Personal and Social Perspectives: Personal health; Populations, resources, and environments; Natural hazards; Risks and benefits; Science and technology in society.

Language Arts Standards

NCTE 4 Students adjust their use of spoken, written, and visual language (e.g., conventions, style, vocabulary) to communicate effectively with a variety of audiences and for different purposes.

NCTE 5 Students employ a wide range of strategies as they write and use different writing process elements appropriately to communicate with different audiences for a variety of purposes.

❖**Materials**: Copies of the story "The Rattlesnake That Tamed a Boy"; chalkboard, overhead, or project the story onto a smart board; pen or pencil and paper; reference works about snakes

Instructional Procedures

Introduction: One of the most important skills of modern science is learning to tell the difference between fact and fiction, such as discerning junk science about the latest diet or incredible claims in a TV commercial from what is real. A tall tale provides plenty of practice in this important life skill. This tale contains much useful information about the environment in general and about snakes in particular. The irony of using a tall tale to teach about the myths about snakes is not lost on most students, after they realize it is a tall tale!

Initiate a discussion about fact versus fiction. Define the difference between fact or truth, and what is false or lies. Ask students to name some ways scientists discover facts and some ways they discover errors. Mention some scientific discoveries that later proved to be in error—or were outright hoaxes.

Activity: After a general discussion of facts and fiction, narrow the focus to snakes.

Make several copies of the story "The Rattlesnake That Tamed a Boy." On the chalkboard or overhead, make a chart with three columns. Label the columns Fact, False, and Do Not Know. Starting at the beginning of "The Rattlesnake That Tamed a Boy," find bits of information—or misinformation—that fit each category. After a few examples, have students work with a partner, in small groups, or as a class to comb the rest of the story for more items to fit in each category.

Students may disagree or simply not know whether certain bits of information are fact or fiction. These pieces of information go in the Do Not Know column. Assign a student or two to sit at a computer and fact check online. Students can research these issues on the spot. Or, for homework, assign each item in the Do Not Know column to a different student. The students research the items online or in the library and report back to the class the next morning. The researchers should report not only their findings, but also how they found and verified the answer. Internet sources also need verifiable fact checking: Was it from DiscoveryChannel.com or some amateur snake collector's online report?

This activity involves higher-level thinking skills, such as inference, deductive thinking, and contrast and comparison. Students evaluate the reliability of information sources and question things that appear to be patently obvious. They are asked whether the things they assume to be true are actually based on fact.

It is important to emphasize during these discussions that opinions are valid only if they are based on facts, not on hearsay or circumstantial evidence. (A discussion of the scientific evidence in a recent national news story may be useful here.) Ask students to defend their answers with the simple questions, "Why do you think this is true?" or, "How do you know?"

Assessment: Ask students to write a very brief, 500-word essay that outlines which parts of the story they believed and where the tangle of lies started to fall apart. What were their first clues that this might be a tall tale? If they believed the

---

whole thing, which often happens, what led them to believe it? What did they learn from this process?

Follow-Up Activities: Students could be asked to collect newspaper or magazine ads that tout noncredible claims, (i.e., weight loss, hair growth, pimple-free, etc.), and then write short research papers testing their claims.

The Liar's Game

❖**Grade Levels**: 2–12 **Time estimate**: 45–60 minutes

❖**Science skills**: Classification; Communication; Prediction; Inference; Identify Variables; Reorder, Analyze, and Draw Conclusions

❖**Objectives**: Students discern fact from falsehood.

National Standards

Science Standards

NAS 1 Science as Inquiry: Abilities necessary to do scientific inquiry; Understandings about scientific inquiry.

NAS 3 Life Science: Structure and function in living systems; Reproduction and heredity; Regulation and behavior; Populations and ecosystems; Diversity and adaptations of organisms.

NAS 5 Science and Technology: Abilities of technological design; Understanding about science and technology; Abilities to distinguish between natural objects and objects made by humans.

NAS 6 Science in Personal and Social Perspectives: Personal health; Populations, resources, and environments; Natural hazards; Risks and benefits; Science and technology in society.

NAS 7 History and Nature of Science: Science as a human endeavor; Nature of science; History of science.

Language Arts Standards

NCTE 1 Students read a wide range of print and nonprint texts to build an understanding of texts, of themselves, and of the cultures of the United States and the world; to acquire new information; to respond to the needs and demands of society and the workplace; and for personal fulfillment. Among these texts are fiction and nonfiction, classic and contemporary works.

NCTE 4 Students adjust their use of spoken, written, and visual language (e.g., conventions, style, vocabulary) to communicate effectively with a variety of audiences and for different purposes.

NCTE 5 Students employ a wide range of strategies as they write and use different writing process elements appropriately to communicate with different audiences for a variety of purposes.

NCTE 7 Students conduct research on issues and interests by generating ideas and questions, and by posing problems. They gather, evaluate, and synthesize data from a variety of sources (e.g., print and nonprint texts, artifacts, people) to communicate their discoveries in ways that suit their purpose and audience.

NCTE 11 Students participate as knowledgeable, reflective, creative, and critical members of a variety of literacy communities.

❖**Materials**: Pen or pencil and paper

Instructional Procedures

Introduction: A popular folk game is the liar's contest. Students take turns sharing three alleged facts about themselves, knowing one is not true, and then the class votes to see if they were fooled. It works best if you begin by sharing three such facts about yourself.

Activity: Students then write a list of three facts about themselves. Two of the facts are indeed true, but one is not. To start the game, one student reads his or her list aloud to the class. The students try to identify which item in the list is a lie. The student who picks out the lie goes next.

Following is an example, using the information presented in "The Rattlesnake That Tamed a Boy." In preparation for the game, read the list to the class and ask students to identify which item is false.

1. My Uncle John has a farm in Tennessee. (True.)

2. I visited my Uncle John's farm when I was in the third grade. (False; I have never visited his farm.)

3. There were rattlesnakes, copperheads, and cottonmouths, three kinds of venomous snakes on my Uncle John's farm. (True.)

The Scientific Twist

To give the game a scientific twist, ask students to list two facts and one error or falsehood about a specific type of plant or animal that correlates with the curriculum. For example, a class studying a certain ecosystem might be restricted to plants and animals in that ecosystem. A class studying a certain phylum or class of plants or animals could be restricted to species from that class, such as reptiles. To prepare for this activity, read the following bits of information about rattlesnakes to the class. Ask students to identify the fact that is not a fact.

1. Rattlesnakes lay eggs. (False. All rattlesnakes give live birth.)

2. You can estimate a rattlesnake's age by counting its rattles. (True, though not exact.)

3. Rattlesnakes hibernate in the winter. (True.)

To expand this activity, students ask their parents and grandparents to tell them about scientific facts or assumptions that have been proven or disproven in their lifetime. Keep in mind that one of the goals is to fool the listener. Tricky statements with unexpected answers are okay. One example is:

1. Swine flu is a disease that originated from a pig flu. (True.)

2. You can get swine flu from eating pork. (False.)

3. Infected humans can give pigs the flu. (True.)

At this point, you may want to discuss facts and myths about swine flu, the avian flu, and other myths that have been proven false.

Assessment: Students should turn in their lists of three facts. These can be graded for grammar, wit, and science content. A second grade or bonus points could be offered based on the percentage of the class they fool.

Follow-Up Activities: I have gotten feedback from a few parents and teachers that this game is bad because it teaches children how to lie, which completely misses the point. The purpose is not to encourage lying but to help students learn to tell when they are being lied to or misled, or when the truth is being distorted. From Obama being a Muslim to supposed news reporters telling us to invest in gold, the media is full of half-truths and outright lies. It is vital that students learn to discern and cultivate the healthy skepticism that is a key science skill: show me the evidence! After playing the liar's game, discuss ways in which this healthy skepticism and the skills of discernment might help scientists.

Bring a Snake to School

❖**Grade level**: K-12

Time: 15 to 30 minute discussion, but it could stay all year!

❖**Science skills**: Observation; Classification; Communication

❖**Materials**: Friendly snake, herpetologist

Snakes are maligned, and there is much misinformation about them. To combat these misunderstandings, bring one or more snakes into your classroom. Most nature centers and zoos have snake handlers who will bring snakes to show your class. Most snake handlers are well prepared to answer students' questions and to entice them with unusual facts and exotic information. Some pet stores may be willing to let you borrow one of their snakes. Or, ask the pet store to put you in touch with a local herpetologist who would be eager to tell your students the truth about these helpful reptiles. With a snake in hand, students are likely to overcome their fears and to retain intriguing facts about these reptiles and their role in the environment.

Snakes make intriguing classroom pets. It is especially exciting to watch them swallow their prey. As with any classroom animal, snakes offer many learning opportunities. Students can research their diet, weight gain, sleeping habits, and other behaviors.

Telling Tales Tall and True

❖**Grade Levels**: 2–12   **Time estimate**: 2 class periods

❖**Science skills**: Classification; Communication; Prediction; Inference

❖**Objectives**: Students will exercise creative writing skills as they walk through the writing process approach.

They will demonstrate their knowledge of the differences between fiction and nonfiction.

National Standards

Science Standards

NAS 1 Science as Inquiry: Abilities necessary to do scientific inquiry; Understandings about scientific inquiry.

NAS 3 Life Science: Structure and function in living systems; Reproduction and heredity; Regulation and behavior; Populations and ecosystems; Diversity and adaptations of organisms.

NAS 6 Science in Personal and Social Perspectives: Personal health; Populations, resources, and environments; Natural hazards; Risks and benefits; Science and technology in society.

Language Arts Standards

NCTE 1 Students read a wide range of print and nonprint texts to build an understanding of texts, of themselves, and of the cultures of the United States and the world; to acquire new information; to respond to the needs and demands of society and the workplace; and for personal fulfillment. Among these texts are fiction and nonfiction, classic and contemporary works.

NCTE 2 Students read a wide range of literature from many periods in many genres to build an understanding of the many dimensions (e.g., philosophical, ethical, aesthetic) of human experience.

NCTE 3 Students apply a wide range of strategies to comprehend, interpret, evaluate, and appreciate texts. They draw on their prior experience, their interactions with other readers and writers, their knowledge of word meaning and of other texts, their word identification strategies, and their understanding of textual features (e.g., sound-letter correspondence, sentence structure, context, graphics).

NCTE 4 Students adjust their use of spoken, written, and visual language (e.g., conventions, style, vocabulary) to communicate effectively with a variety of audiences and for different purposes.

NCTE 5 Students employ a wide range of strategies as they write and use different writing process elements appropriately to communicate with different audiences for a variety of purposes.

NCTE 6 Students apply knowledge of language structure, language conventions (e.g., spelling and punctuation), media techniques, figurative language, and genre to create, critique, and discuss print and nonprint texts.

NCTE 12 Students use spoken, written, and visual language to accomplish their own purposes (e.g., for learning, enjoyment, persuasion, and the exchange of information).

❖**Materials**: Telling Tales Tall and True worksheet; anthologies of tall tales; pen or pencil and paper

Instructional Procedures

Introduction: Initiate a discussion about what sets tall tales apart from other kinds of stories. Encourage students to think about the style and content of tall tales, as well as their origin and purpose. Ask students to compare "The Rattlesnake That Tamed a Boy" to other stories or books they have read recently. How are they alike? How are they different? What makes one a tall tale and another merely a work of fiction?

Activity: Read or tell several tall tales to the class. Or, better yet, challenge students to research and rehearse tall tales to present to the class. Choose tall tales that are suitable for your area. Every region of the country has its own set of tall tales. In states where lumbering is big business, tales of Paul Bunyan abound. Along the Mississippi it is Annie Christmas that folks talk about. In the southwest, many *cuentos*, or Hispanic folk stories, are tall tales. Ask students to identify some of the characteristics those tall tales have in common. Explore why tall tales came into being: What purpose do they serve? Though they are clearly fiction, they are often created to try to explain natural phenomenon. In this way they are like science!

Encourage students to identify the two types of tall tales: those that start with a lie that grows bigger, and those that start with the truth and then stretch it until it breaks. Then distribute the worksheet Telling Tales Tall and True.

Use the template of the handout and walk students through the process of writing their own tall tales. They can decide if they want to start with the truth and stretch it, like the pet rattlesnake, or start with a lie and grow a whopper, like Pecos Bill.

Assessment: These stories can be collected and graded like any other Language Arts assignment.

Follow-Up Activities: Students can edit and publish their stories online, in a classroom collection, or as a young authors' book. They could also perform them for the class or for other classes.

Telling Tales Tall and True

Mark Twain once said, "Though there is truth in every story, not every story is true." Decide which parts of "The Rattlesnake that Tamed a Boy" are true and not true. Make a chart. Make three columns: one for the truth, one for the lies, and one for the parts that you do not know. Later, you could research the parts you do not know and add them to one of the other columns.

THE TRUTH	THE LIES	I DO NOT KNOW

Retell a story you heard today, but change some of the names and places so it sounds like it really happened to you!

There are two kinds of tall tales: some begin with a lie and only get bigger; others start with the truth and slowly stretch it until it breaks. Let's take a closer look at how you create both types of tales.

The tall tales that start with a lie are often stories about people like Paul Bunyan and Pecos Bill, John Henry, and Annie Oakley. Check your library for more stories about these characters. These may be real people, but that is where the truth ends.

To write a story like this:

1. Start by describing a superhuman character who can perform amazing feats of strength and agility. Write two or three sentences about your character. Describe both the person's physical appearance and what kind of person he or she is.
2. Write two sentences describing the setting for your story, including both time and place.
3. Write one sentence describing some fantastic problem.
4. Write two sentences describing how ordinary people try to solve the problem and why they fail.
5. Write two sentences describing how the hero saves the day using his or her special powers.

These sentences together add up to an outline of your story. Reshape these sentences into a story. Rewrite, revise, and edit these stories, and then collect them into a class book. Take turns telling these stories to your class.

♠ ♣ ♦ ♥ Though there is truth in every story, not every story is true! ♠ ♣ ♦ ♥

Tall tales that start with the truth are my favorite stories to tell because it is a lot of fun to watch the audience as they get tangled into the net of lies—but still believe the story. These are the hardest stories to write because it is difficult to get the listener to believe you and then maintain that trust as you stretch the truth to its breaking point.

It helps if you start with a true event from your life that is very believable yet somewhat unusual. Think back to some amazing thing that really did happen to you. Close your eyes and relive it in your memory. Use all of your senses to make the memory come to life. Write several sentences describing the event, the people, and the setting. How can you stretch it into a tall tale? Start with little facts: exaggerate the numbers, talk about believable things that could, but did not happen. Pull on it a little more. Add information that is barely possible but could still happen. Finally, wrap it up with things that are impossible. I like to end with a line like: "It was just lying there like I have been lying to you the whole time," or "She was yanking on my leg just like I am pulling your leg now." If you gently bring the listener along they will take it, hook, line, and sinker!

For added fun you could have a tall tale contest to see how many people you could fool. Tell your story to your class and see if they believe you. Do a survey of your class to see how many you fooled.

Counsel of Crows

The Wisdom of Our Elders: Human and Otherwise!

Dedicated to Grandfather Warren and my buddy, Thunderhorse

He told me how a friend of his once heard a whole sky full of stars when she was seven. And later on when she was eighty-three she heard a cactus blooming in the dark.

—Byrd Baylor, from *The Other Way to Listen*

Comments to the Teacher

THIS STORY EXPLORES SEVERAL IMPORTANT IDEAS: What can we learn from the explanations of the natural world given by Aboriginal elders? What can we learn from the inherent ecological intelligence in cultural folktales? What can we learn from the behavior and wisdom of other species? How is the intelligence of animals different from the intelligence of humans?

This story assumes that animals are intelligent *and* that various cultures have different but equally valid perspectives on nature and science. It assumes that the traditional folktales of many cultures embody wisdom about our relationship with nature, and it also assumes that animals have intelligence, different from ours, that allows them to flourish. With these assumptions in place, the core question is: What can we learn from various cultures and other species? Although this story is anthropomorphic, that technique is used to challenge the listener to be open to the wisdom of other species.

And now the story . . .

One sunny day, just before the summer solstice, I helped a friend of mine who went on a vision quest. Now, a  vision quest is a ceremony for Native American boys and girls, young men and young women, to help them gain a deeper sense of themselves, to figure out who they are, what their gifts or strengths are, and what they can contribute to their tribe. A vision quest can also be performed when you face an important transition in life, for example, from teenager to adult, or from adult to elder. Though the ceremony differs among Native American tribes, generally you go off alone for four days and four nights with no food. You fast and you remain alone, so you can become closer to the spirits of nature, to the Creator. You watch for a sign or a communication from the natural world. You cry for a vision. You try to answer questions like, "Who am I? Why am I here? What can I do to make the world a better place?"

Well, a young friend of mine asked me if I would tend fire for him. To tend fire is  to be his assistant and his coach. When he went off on a hill by himself, I stayed at the fire to make sure that he was safe. If he had any problems, I would be there 24 hours a day to help him out. Also, when he came off the hill, after his vision quest, part of my duty was to help him to understand his vision and to assist him in making his dreams come true.

While I was sitting by the fire, a Native American elder, Grandfather Warren, came to visit. He is a Choctaw/Chickasaw elder and friend of the young man on the hill. Grandfather Warren came and camped out with me for a couple of days.

Right away he began to ask me all kinds of questions, like, "Why did you do this that way? Why did you do that this way?" I thought he was testing me, to see if I was up for the challenge that faced us. But he said he was not. He said he questioned me because my way of doing things was different from his and he wanted to understand me better. After we had built a foundation of trust, Grandfather Warren began to tell stories about his youth, his training and his elders. While he was sharing his stories . . . *Caw, caw, caw!* We heard a flock of crows fly out of the forest.

This was a huge flock of crows. Have you seen a flock of crows fly? Isn't it beautiful the way one crow turns, and all the other crows turn at the same time? They never run into each other. It's as if they fly with one mind, one body.

The crows flew out of the forest where my friend was camping. Grandfather Warren and I watched them to see if they would give us a hint as to what was going on in the forest.

Can you imagine being a bird? Close your eyes and imagine that you are a crow. Imagine that you are covered with feathers, dark black feathers that glisten with a rainbow of colors. You have black feet and a black beak. Open your eyes and stretch out your wings. Don't bump into anybody! Imagine you are flying. Feel the wind in your feathers as you lift and curve. Feel the wind as you lift up into the sky, as you curve into the forest. As you land on a tree branch, grab it with your feet and fold back your wings.

That's what I did while I watched the crows fly. I imagined that I landed in the tree with the crows. The crows looked at me as if to say, "What are you doing here? You're not a crow." Well, I have learned when to be quiet, and when to talk, when to ask questions and when to listen. I was quiet.

One of the crows said, "Caw, caw, caw! You two-legged creatures think you know everything. But you could learn much from the other animals, from the winged, from the four-legged, from the creepy crawlies. If you had ears to listen and eyes to see. Caw, caw! You could learn much from a crow."

Then a younger crow, a baby crow, chimed in. "Caw! In our flock we watch out for each other. We care for every young crow. Once, an owl attacked me in my nest. The owl tried to eat me! But the crows ganged up on that owl. One owl could eat one crow, but when 10 crows attacked, they chased the owl away. Caw, caw, caw! They saved my life. Do you watch out for your friends? Do you help one another?"

Then another young crow spoke.

"Caw. You look at me and you see black. You look only at what is obvious. Look past the surface, look inside to what is true. Have you ever seen a crow feather? Hold a crow feather up to the

sunlight, and you will see a rainbow of colors, blue and green and purple. Look beyond the obvious and you will see the rainbow! Caw!"

Those crows really challenged me to think. As I sat up there in the tree with them, I began to understand them better, and I began to remember other things I had learned about crows.

Do you know Aesop's fables? Aesop told many fables about crows. One of my favorite fables is about the crow and the pitcher of water. Once there was a pitcher that was half filled with water. Along came a crow who was very thirsty. But the pitcher had a very narrow neck, and the crow could not get her head down far enough into the pitcher to drink the water. Now, some people might have given up, but not this crow! The crow used her brain to solve the problem. She found some pebbles and dropped them into the pitcher until the water rose to the rim of the pitcher. Then she drank her fill. Recently, a scientist replicated this story as an experiment and found that crows are very good with this kind of problem solving. A real crow placed pebbles into a pitcher to get a drink. This was no fairy tale![1]

Yes, I remembered those old fables of Aesop's, and he was right. Crows are very smart. As I watched them in the tree, talking among themselves, I remembered reading a report about a scientist who studied crows to see if they could communicate. He found that crows can learn a variety of human symbols, they can count, add, subtract, and they are excellent problem solvers. I had to chuckle at the thought that the birds could understand the scientist's language, but he could not understand theirs! I listened carefully to the crows nearby. I could understand them when they spoke. But then, I was sitting in their tree, and I felt as though they were training me!

By human standards that scientist considered crows intelligent. But his tests did not even begin to measure the intelligence they need to survive. He never learned how they migrate using the stars, how their series of squawks and caws convey meaning, how they band together in flocks with respect for elders and compassion for their young. There was much about the crows' world the scientist couldn't measure or quantify!

Well, I got to thinking so hard about this that I forgot where I was sitting. I started to slip off balance on my branch without even noticing. Just before I lost my balance and fell, a crow nipped me on the seat to wake me up.

"Caw!" it called. "Look at this one, falling out of a tree! And they call us birdbrains! That should be considered a compliment."

Then the oldest crow spoke. Her feathers had begun to turn white, so you could tell she was very old. She said, "Caw, caw. The Shawnee, who have lived in this valley since before you were here, they tell a story about the beginning of the world. When the world was new, Creator made the world with her every thought. She thought about the sun, and the sun was made. She thought about the moon, and the moon was made." The old crow looked right at me. "You might think that's strange, but think about something you have made. Did you first imagine it? Whatever you can imagine you can create. Caw.

"But Creator had to be careful, because if Creator thought an evil thought, then evil would be made.

"One day, Creator spent the whole day making creepy crawlies. She worked from sunup until sundown, making the most amazing variety of insects and spiders, with an astounding variety of colors and shapes. After creating all those creepy crawlies, Creator was tired. Because she was so tired, Creator was not careful with her thoughts. Then an evil thought entered her mind. Oh no! Evil was made, evil was let loose in the world! She tried to stop it, but it was too late.

"But Evil was not a complete thought; Evil was made without a head. So Evil went looking for a head. Evil saw a natural spring where water came gurgling out of the rocks. Evil noticed that the animals came there to drink. Evil hid in the bushes and waited. Along came a huge deer, with 14 points in its huge antlers. Evil thought, Ah, that's the head I want! Evil hid in ambush. But Evil did not count on the crow sitting in a tree nearby. When the deer bent over to drink, the crow saw what was coming. The crow called out, *Caw, caw, caw!*

"The deer jumped, Evil fell into the water—*splash!*—and the deer ran away. As it ran, the deer waved its white tail to warn the other deer.

"That is why, to this day, the Shawnee do not go into the forest without a crow feather or hair from the white tail of a deer sewn into their clothes or braided into their hair to ward off Evil. To this day, Evil is out there looking for a head. Evil cannot act without one of us giving it our head. And to this day the crow is the guardian of the forest. Caw, caw."

Then the first crow chimed in again.

"Caw, caw, caw! You two-legged creatures think you know everything. If you had ears to listen and eyes to see, you could learn much from the creatures of the forest. You could learn much from a crow! Caw!"

Then the crows took off in flight.

They were so beautiful, twisting and turning, their feathers reflecting the colors of the rainbow. But I was a little bit nervous, left in the tree in the forest all by myself. That talk about Evil made me nervous, and I wished I had just one feather from one of those crows.

Then the old crow circled round and came back to land on the branch above me. "You do not have to worry," she said. "Unless you think evil thoughts, evil cannot hurt you. Be careful what you think." Then she took off, and one of her feathers floated down. As it floated past me, I caught it and pinned it to my cap. All the way back to the campfire, I thought about what the crows had said, about being careful what you think, about listening, and about taking care of each other.

I do not know about you, but I, I felt honored to sit among a council of crows.

That was my vision. I came to the forest to tend fire for my friend, and I went away with a new understanding of the varieties of wisdom on our planet. As for my friend, he had an invigorating experience, but what his vision was I cannot tell. That is another story; that is his tale to tell.

Follow-Up Ideas for "Counsel of Crows"

Discussion Topics

Immediately after reading this story, ask students: What did this author learn from the counsel of crows? What did you learn? Ask students to share stories of close encounters with wild animals. Not just, "I saw a deer," but some special moment where it felt like the animal was communicating with you. What did you learn from this creature?

Porque (What-For) Stories: Folktales as a Source of Scientific Wisdom

❖ **Grade Levels**: K–12   **Time estimate**: 50–60 minutes

❖ **Science skills**: Communication; Inference; Formulate Hypotheses

❖ **Objectives**: Students will explain natural phenomenon with inventive stories.

They will demonstrate their understanding of animal behavior and habitats.

National Standards

Science Standards

NAS 1 Science as Inquiry: Abilities necessary to do scientific inquiry; Understandings about scientific inquiry.

NAS 3 Life Science: Structure and function in living systems; Reproduction and heredity; Regulation and behavior; Populations and ecosystems; Diversity and adaptations of organisms.

NAS 6 Science in Personal and Social Perspectives: Personal health; Populations, resources, and environments; Natural hazards; Risks and benefits; Science and technology in society.

NAS 7 History and Nature of Science: Science as a human endeavor; Nature of science; History of science

Language Arts Standards

NCTE 1 Students read a wide range of print and nonprint texts to build an understanding of texts, of themselves, and of the cultures of the United States and the world; to acquire new information; to respond to the needs and demands of society and the workplace; and for personal fulfillment. Among these texts are fiction and nonfiction, classic and contemporary works.

NCTE 2 Students read a wide range of literature from many periods in many genres to build an understanding of the many dimensions (e.g., philosophical, ethical, aesthetic) of human experience.

NCTE 4 Students adjust their use of spoken, written, and visual language (e.g., conventions, style, vocabulary) to communicate effectively with a variety of audiences and for different purposes.

NCTE 5 Students employ a wide range of strategies as they write and use different writing process elements appropriately to communicate with different audiences for a variety of purposes.

NCTE 7 Students conduct research on issues and interests by generating ideas and questions, and by posing problems. They gather, evaluate, and synthesize data from a variety of sources (e.g., print and nonprint texts, artifacts, people) to communicate their discoveries in ways that suit their purpose and audience.

NCTE 9 Students develop an understanding of and respect for diversity in language use, patterns, and dialects across cultures, ethnic groups, geographic regions, and social roles.

NCTE 11 Students participate as knowledgeable, reflective, creative, and critical members of a variety of literacy communities.

❖**Materials**: One or more books of *porque* stories to read aloud; copies of the handouts "How Fox Earned His Tail" and "Plans for My *Porque* Story"; pen or pencil and paper

Instructional Procedures

Introduction: "Counsel of Crows" raises an important question about the wisdom of other cultures, which early scientists dismissed as superstition. If you look beneath the surface of traditional folktales, you will find a vast amount of information about the ecosystems of indigenous peoples. In a simple *porque* story you can learn which plants and animals live in an area, learn about predator–prey relationships, and learn about the habitats and habits of local animals. Usually, this information is not the point of the *porque* story; the point is to highlight human foibles and concerns. But the information is there, nevertheless.

Many native cultures the world over have a complex understanding of herbal medicines, fruiting schedules, animal behavior, solar and lunar calendars, rain cycles, and other important information about their connection to the natural cycles. Indigenous people strive to live humbly within these rhythms, rather than impose their order on things. Since the beginning of time, humans have used stories and myths to explain their world, to answer the who, what, when, why, where, and how of existence. Seen in this light, as a way to explain the natural world, science could be considered the cosmology of the modern world.

Activity: Read aloud or tell a few of your favorite *porque* stories. (Please see "How Fox Earned his Bushy Tail" on page 125.) There are many wonderful examples of this type of story in cultures throughout the world.

After telling the stories, ask students to turn to a partner and discuss whether the story is true, or which parts are true and which are not. Why or why not? As a class, discuss the scientific reasons why chipmunks have stripes or elephants have long trunks or why mosquitoes buzz in people's ears.

Pass out the worksheet on writing *porque* stories on page 00. Use this template to help students write a story about some important transformation in their favorite animal. Emphasize the importance of including an accurate description of the animal's habitat and behavior.

Some possible topics include:

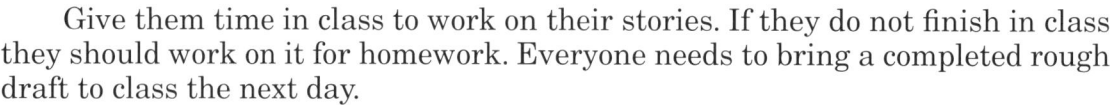

- How did porcupines get their spines?

- How did birds get their wings?

- Why do giraffes have long necks?

- Why do beavers have wide tails?

- Why do humans stand upright?

Give them time in class to work on their stories. If they do not finish in class they should work on it for homework. Everyone needs to bring a completed rough draft to class the next day.

On the second day, challenge students to review and rewrite the scientific details of the story. Remind them that they can get two grades for this assignment, one for language arts and one for science, but if their science is way off they can lose points! Allow them a few minutes to share their rough draft with a partner. The partner's job is to listen and ask questions, not to criticize, but to coach and encourage. Each of them can help their partner flesh out some of the detail. Allow them the rest of the period to finish rewriting and editing their final version of the story.

Assessment: Collect the worksheet and their rough draft of their story, as well as the final draft of the story. Evaluate the mechanics of their writing as well as their science content.

Follow-Up Activities: Allow students to illustrate their stories and perform them in small groups or for the class. After writing their stories, allow students to research the scientific explanation for their topic.

Here is one of my favorite *porque* stories, written with the help of a fifth-grade class in Lansing, Illinois:

How the Red Fox Earned His Bushy Tail

We have all heard stories of the animals who have lost their tails: how rabbit fell from a tree and his tail snapped off, creating the puff on a pussy-willow; how bear tried to catch a fish with his tail and it froze in the lake; how possum's vanity led cricket to clip his thick fur. But here is a story of valor and self-sacrifice, the story of how the red fox *earned* his bushy tail.

Long, long ago the red fox did not look as he does today, with his bright red fur, black feet, and beautiful bushy tail. His fur was a drab, spotty gray and his tail was skinny and hairless, more like a rat's tail than a squirrel's. Oh, he has always been clever and sometimes self-absorbed, but we will get to that later.

One day the fox was hungry and decided to go hunting. Fox was always hungry in part because he was not a very good hunter. Fox was fast, faster than rabbit, but rabbit could dodge and turn and often got away. Fox was more clever than mouse, but his drab fur was not the soft red-orange-brown of dried leaves. His distinctive color meant that mouse could always see him coming and slip quietly into her snug little den.

After another unsuccessful hunt, fox sat by the side of a pool of water looking at his reflection. He thought to himself, "I am so drab. I am ugly. I am hungry. It's not fair. I am faster than rabbit but I cannot catch her. Why is it that mouse always sees me coming? Why am I always so hungry? Why am I always cold? I wish my fur would blend in better. I wish I could dodge and turn as quick as a rabbit." On and on the fox grumbled and mumbled to himself.

As he sat there feeling blue, dwelling in self-pity, drowning in his misery, he smelled something. He smelled smoke. He thought he heard crackling. Yes, he heard the crackling of fire. Looking out across the meadow he saw a fire raging through the field. It was being blown toward the forest, toward his home.

At first he was afraid. "Oh, no! My home will be burned! My wife and children will be scorched!" Then he thought about all the other animals whose homes would be ruined, and all the lives that would be lost. Even his nemesis the rabbit might lose her life: her resting-place under the bush was already up in flames.

As he began to think about something besides himself, as he began to think about his family and the other animals of the forest, a new feeling came over him. He was no longer afraid. He knew a new kind of bravery. He knew he had to act.

Quickly, he ran out in front of the fire. He began stomping on the flames. The fox shrieked and shouted as loud as he could, "Run! Run for your lives! There is a fire in the meadow and it's burning towards the forest!" As he shouted, he continued stomping on the fire. It was not working. His feet were not big enough, but he was clever and his heart was growing to match the quickness of mind. The fox dropped to his side and began rolling across the fire, extinguishing the flames.

From *Learning from the Land: Teaching Ecology through Stories and Activities,* Second Edition by Brian "Fox" Ellis. Santa Barbara, CA: Libraries Unlimited. Copyright © 2012.

The animals heard his cries and all of them escaped. Their homes were ruined for now, but they knew the forests and prairies renew themselves. The flames actually enrich the soil; the ash is fertilizer and you must clear away the old plants to make room for new growth. When the flames had died down and the animals returned to assess the damage, they were stunned to see the fox lying there scorched and dead.

Or was he? As the creatures of the meadow gathered around they saw him slowly take in a breath. It was actually the mouse who said, "Look, he's still alive! I felt the air come out of him."

As the animals stood in silence, staring, the rabbit whispered a prayer. The rabbit knew that it was the fox who kept her fast, it was the fox who kept her alert, and as her kin fed the fox, the fox kept the warren healthy and strong by eating the old and sick. With these thoughts in her heart she whispered a prayer: "Please, Creator, save him as he has saved us."

Just then a gentle breeze blew a whirlwind of dust and ash that seemed to sparkle in the sunlight. Before the astonished eyes of all the creatures the fox began to sprout new fur, not the drab gray he grew before, but a fur that was a brilliant orange and red like the flames they had seen just moments before. His feet and legs sprouted black fur like the ashes fox had stomped so bravely. And then he grew thick fur on his tail to help him dodge and turn and to keep him warm on a cold night. His wish had been granted.

Soon a healing rain began to fall across the forest and fields. New life sprouted everywhere. The mouse and rabbit quietly slipped away, knowing they would be chased by a better hunter. The fox slowly stood up. He looked at himself in the nearest puddle and knew that these new gifts were given because of his valor.

To this day fox has a bushy tail to help him dodge and turn, to keep him warm on a cold winter's night. He has red-orange fur to help him blend in well with the dried leaves of the forest and the burnt orange grasses of the prairie. And to this day the red fox has scorched black feet to help him remember this story, to remember the time he risked his life to help those around him.

Plans for My *Porque* Story

Name_____

The type of animal I have chosen: _____.

Write a brief description of its unique characteristics. How does it look today?

_____
_____
_____

Write a brief description of how it once was before it changed. How did it change?

_____
_____
_____

Write a brief description of its habitat or home including information about weather, food, shelter, landforms, and other plants and animals who share its home.

_____
_____
_____

What problem does your animal face? How does he/she solve it?

_____
_____
_____

Make an outline of your story:
Beginning:

_____
_____
_____

Middle:

_____
_____
_____

End:

_____
_____
_____

What does your main character learn from this adventure? What can we learn from his/her mistakes and successes? What is the moral of the story?

_____
_____
_____

Use this information as an outline for writing your own how and why story!

From *Learning from the Land: Teaching Ecology through Stories and Activities,* Second Edition by Brian "Fox" Ellis. Santa Barbara, CA: Libraries Unlimited. Copyright © 2012.

Animal Wisdom

❖**Grade Levels**: 3–12 **Time estimate**: Two class periods with some homework

❖**Science skills**: Observation; Communication; Inference; Reorder, Analyze, and Draw Conclusions; Design Investigations

❖**Objectives**: Students will demonstrate both a scientific and cultural interpretation of an animal's wisdom.

Students will learn research skills and how to translate their notes into a story.

National Standards

Science Standards

NAS 1 Science as Inquiry: Abilities necessary to do scientific inquiry; Understandings about scientific inquiry.

NAS 3 Life Science: Structure and function in living systems; Reproduction and heredity; Regulation and behavior; Populations and ecosystems; Diversity and adaptations of organisms.

NAS 7 History and Nature of Science: Science as a human endeavor; Nature of science; History of science

Language Arts Standards

NCTE 1 Students read a wide range of print and nonprint texts to build an understanding of texts, of themselves, and of the cultures of the United States and the world; to acquire new information; to respond to the needs and demands of society and the workplace; and for personal fulfillment. Among these texts are fiction and nonfiction, classic and contemporary works.

NCTE 2 Students read a wide range of literature from many periods in many genres to build an understanding of the many dimensions (e.g., philosophical, ethical, aesthetic) of human experience.

NCTE 4 Students adjust their use of spoken, written, and visual language (e.g., conventions, style, vocabulary) to communicate effectively with a variety of audiences and for different purposes.

NCTE 5 Students employ a wide range of strategies as they write and use different writing process elements appropriately to communicate with different audiences for a variety of purposes.

NCTE 7 Students conduct research on issues and interests by generating ideas and questions, and by posing problems. They gather, evaluate, and synthesize data from a variety of sources (e.g., print and nonprint texts, artifacts, people) to communicate their discoveries in ways that suit their purpose and audience.

NCTE 8 Students use a variety of technological and information resources (e.g., libraries, databases, computer networks, video) to gather and synthesize information and to create and communicate knowledge.

NCTE 9 Students develop an understanding of and respect for diversity in language use, patterns, and dialects across cultures, ethnic groups, geographic regions, and social roles.

❖**Materials**: Pen or pencil and paper; access to library research materials, including science books, science journals, CDs, DVDs, and the Internet

Instructional Procedures

Introduction: After hearing "Council of Crows," engage students in a conversation about what animals might need to know that is different from what we might need to know. What is the animal's intelligence? Challenge them to think about Aboriginal people who have lived with these animals for thousands of years and what we might learn from them. What folktales do students know about animals? What culture tells this story? What can we learn about the animal from the folktale?

Activity: Ask students to choose their favorite animal and write its name at the top of a piece of paper. Ask them to brainstorm on paper a list of facts about the animal. To stimulate their ideas, conduct this prior-knowledge quiz. How big is it? What color(s) does it have? What size is it? What is its shape? Does it have fur, feathers, or scales? What does it eat? What eats it? Where does it live? What plants and animals share its home range? What does it do? What are its strengths? Ask these questions in rapid-fire succession. Instruct students to answer as many as they can with short phrases or single words and then to go back and fill in the details.

Finally, have students make a list of questions about the animal. Challenge students to create thoughtful questions, such as, what does this animal symbolize for the people who know it best? What scientific information can I learn from this folk material?

With these notes in hand, challenge students to design their own plan for answering their questions. They could go to the library; they could use books, scientific journals, CDs, videos, the Internet, or any other source. They could visit a zoo or natural history museum to see the creature or to consult a zoologist. They could look for folktales about the animal or study the folk art of various cultures to answer their questions.

Depending on the grade level and research abilities of your students, you may ask them to list 5, 10, or 25 unusual facts about their animal.

After the research is complete, invite the students to use what they have learned to write a story, poem, or song modeled after "Counsel of Crows." They could begin with a dreamlike encounter with the creature, explore the habits and lifestyle of the creature, and end with an emphasis on the animal's role in nature. The undercurrent or theme of their stories is the lessons we could learn if only we could open our minds to the wisdom of other creatures.

Assessment: Students should be forewarned that their research notes will be collected and graded based both on the quality of their questions and on how well they were answered. Maybe bonus points could be offered for using multiple sources: books, Internet, field search, and so forth. Their final stories can also be given two grades based on grammar and science.

Follow-Up Activities: Of all of the stories I am hoping students will write as a result of the lessons in this book, these are the ones I would be most interested in reading. Please forward a copy of the best books to my website! www.foxtalesint. com.

Listening to the Voice of Nature

❖**Grade Levels**: K–12 **Time estimate**: 30–60 minutes

❖**Science skills**: Observation; Communication; Prediction; Inference; Formulate Hypotheses; Reorder, Analyze, and Draw Conclusions

❖**Objectives**: Students will learn to use all of their senses to observe the wild world around them.

They will demonstrate an ability to make inferences and draw conclusions based on this kind of quiet observation.

National Standards
Science Standards

NAS 1 Science as Inquiry: Abilities necessary to do scientific inquiry; Understandings about scientific inquiry.

NAS 7 History and Nature of Science: Science as a human endeavor; Nature of science; History of science

Language Arts Standards

NCTE 4 Students adjust their use of spoken, written, and visual language (e.g., conventions, style, vocabulary) to communicate effectively with a variety of audiences and for different purposes.

NCTE 7 Students conduct research on issues and interests by generating ideas and questions, and by posing problems. They gather, evaluate, and synthesize data from a variety of sources (e.g., print and nonprint texts, artifacts, people) to communicate their discoveries in ways that suit their purpose and audience.

NCTE 11 Students participate as knowledgeable, reflective, creative, and critical members of a variety of literacy communities.

NCTE 12 Students use spoken, written, and visual language to accomplish their own purposes (e.g., for learning, enjoyment, persuasion, and the exchange of information).

❖**Materials**: Pen or pencil and paper, access to a quiet outdoor environment.

Instructional Procedures

Introduction: Native cultures around the world place a high value on learning to listen to the voices of the wild world. Challenge students to spend time quietly listening to the sounds of nature. This activity can be made more or less challenging by asking questions that give the exercise a particular focus.

Activity: You could begin by asking students to sit quietly in a circle for one minute. Ask them to silently count how many sounds they hear. Then discuss the number of sounds and help students to discern the calls of various birds or to tell the difference between the sound of wind in a pine tree and wind in an oak tree.

Ask students to find a quiet place where they can sit silently for three to five minutes. As they sit and listen to the variety of sounds, they ask themselves: What is one idea I could learn from nature?

Finally, allow students to sit alone and silently in a quiet natural environment with a journal. Ask them to record their observations with any of the following questions in mind:

- Using all of my senses, what do I see, hear, smell, feel, notice?

- What is happening here and why does it happen that way?

- Based on what I see, hear, smell, notice, what do I think will happen next?

- What did this place look like 100 years ago?

- What will it look like 100 years from now?

- What can I learn about being a better person from the plants and animals around me?

It may be helpful to do this activity in several sessions, asking students to reflect on one or two questions per session.

Assessment: Yes, this is one of those activities that allows students to simply enjoy themselves in the wild world and see what they can learn in a loosely structured format!

Follow-Up Activities: Students may weave notes from their journals into a poem or short story. This is also the kind of activity that could lead to genuine, student-led inquiry, an inspiration that ignites a lifelong curiosity about the environment.

Native American Tales: Placing Stories in their Eco-Regions

❖**Grade Levels**: 3–8 **Time estimate**: Two class periods

❖**Science skills**: Observation; Classification

❖**Objectives**: Students will make connections between folktales, cultures, and the environment in which they blossomed.

Materials: Books of Native American tales; Native American Tales worksheet; pen or pencil and paper

Instructional Procedures

Introduction: Tell several Native American tales to the class.

Activity: Hand out the Native American Tales worksheet. Get the class started on filling in the sheet with stars placed on the map where each story comes from and the culture and environment that created the story. Ask them to turn to a partner and discuss how the ecology is reflected in the tales told. Introduce the concept of a color-coded key. Allow students to work individually to color-code the sheet.

Assessment: The sheets can be collected and graded for accuracy. Does the color-coding match the key they created?

Follow-Up Activities: This activity is a lead into the next, student storytelling.

Student Storytelling

Grade Levels: K–12 **Time estimate**: Two class periods

Science skills: Communication; Inference

Objectives: Students will research Native American folktales.

They will demonstrate their ability to tell traditional folktales.

National Standards

Science Standards

NAS 1 Science as Inquiry: Abilities necessary to do scientific inquiry; Understandings about scientific inquiry.

NAS 3 Life Science: Structure and function in living systems; Reproduction and heredity; Regulation and behavior; Populations and ecosystems; Diversity and adaptations of organisms.

NAS 7 History and Nature of Science: Science as a human endeavor; Nature of science; History of science.

Language Arts Standards

NCTE 1 Students read a wide range of print and nonprint texts to build an understanding of texts, of themselves, and of the cultures of the United States and the world; to acquire new information; to respond to the needs and demands of society and the workplace; and for personal fulfillment. Among these texts are fiction and nonfiction, classic and contemporary works.

NCTE 2 Students read a wide range of literature from many periods in many genres to build an understanding of the many dimensions (e.g., philosophical, ethical, aesthetic) of human experience.

NCTE 3 Students apply a wide range of strategies to comprehend, interpret, evaluate, and appreciate texts. They draw on their prior experience, their interactions with other readers and writers, their knowledge of word meaning and of other texts, their word identification strategies, and their understanding of textual features (e.g., sound-letter correspondence, sentence structure, context, graphics).

NCTE 4 Students adjust their use of spoken, written, and visual language (e.g., conventions, style, vocabulary) to communicate effectively with a variety of audiences and for different purposes.

NCTE 5 Students employ a wide range of strategies as they write and use different writing process elements appropriately to communicate with different audiences for a variety of purposes.

NCTE 6 Students apply knowledge of language structure, language conventions (e.g., spelling and punctuation), media techniques, figurative language, and genre to create, critique, and discuss print and nonprint texts.

NCTE 8 Students use a variety of technological and information resources (e.g., libraries, databases, computer networks, video) to gather and synthesize information and to create and communicate knowledge.

NCTE 9 Students develop an understanding of and respect for diversity in language use, patterns, and dialects across cultures, ethnic groups, geographic regions, and social roles.

❖**Materials**: Collections of Native American Folktales

Instructional Procedures

Introduction: Students work independently on this project to research, learn, practice, and tell a Native American tale to their classmates.

Activity: Ask students to find out which Native American tribes live in your part of the country. Ask the school library media specialist to do a book talk on Native American tales, and instruct students to choose a tale from a local tribe. Encourage students to try to see the region as it was when the story took place.

Remind students of how to prepare to tell a story. (See instructions in introduction.) Tell them to first read the story several times until they think they know it well. Then ask them: How might you change your voice while telling it out loud? How might you use your body if you were to tell it silently? Encourage students to close their eyes and see the story in their imagination. Remind them to use their voice, body, and imagination together to tell, not read, the story.

Working independently, allow students to practice telling the story to themselves; to their parents; to their friends—even their goldfish, if they have one! Impress upon them the importance of practice, practice, practice. Do not force students to tell their story to the class before they are ready. Allow them ample time to prepare their story, and then arrange a sign-up sheet so they can choose when they are ready to tell their story to the class.

Assessment: For a detailed rubric on evaluating student storytellers, please see my book, *Content Area Reading, Writing, and Storytelling* (Teachers Ideas Press, 2009).

Follow-Up Activities: Encourage students to take their show on the road! Schedule time for them to visit other classes and share their stories. Schedule a family literacy night and let a few of the more enthusiastic entertain the families who attend.

Turtle Island Tales

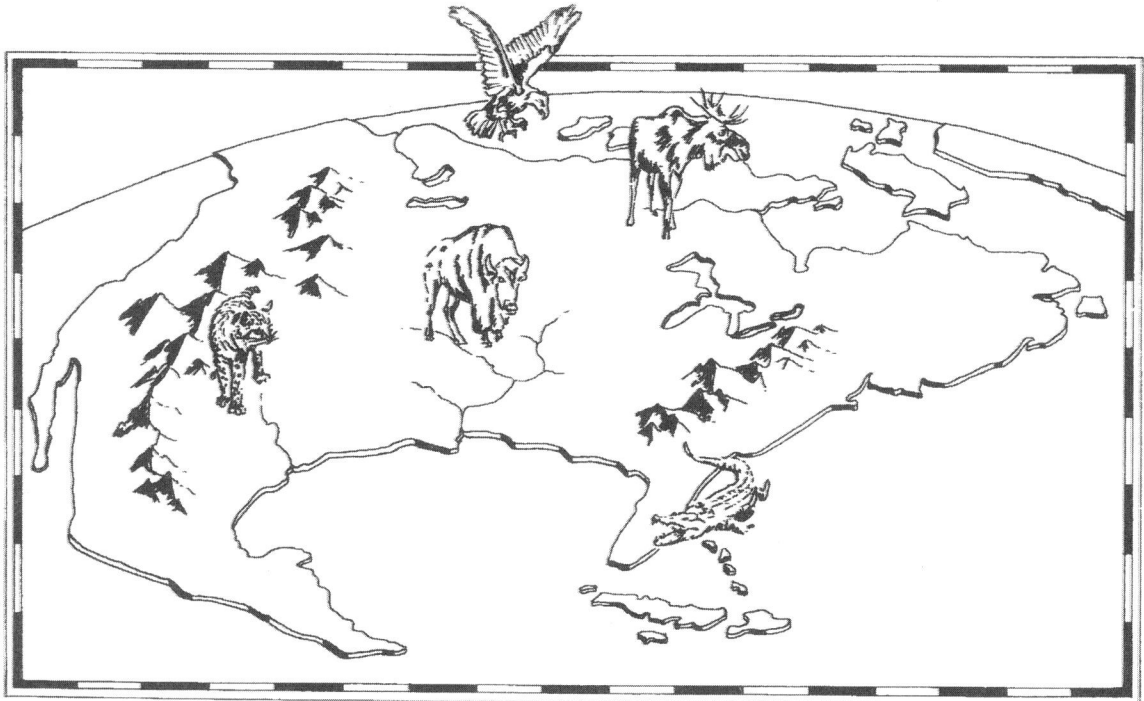

Use this map to find the places where the stories you heard took place. Put a star on the map in these areas. Create a key to color-code the lakes, rivers, mountains, deserts, grasslands, and forests.

Retell a story you heard today as if you were there when it happened!

Find out what tribes live in your part of the country. Go to the library and find a book of American Indian stories. Learn a story from one of the local tribes. As you read the story, try to see your neighborhood as it was when the story took place. Read the story several times until you think you know it fairly well. How might you change your voice while telling it out loud? How might you use your body if you had to tell it silently? Close your eyes and see the story in your imagination. Use your voice, body, and imagination together to *tell*, not read, your story. Practice telling it to yourself, a mirror, your mom, your friend, and your goldfish. Practice, practice, *practice*. When you think you have it the way you want it, ask your teacher if you could tell your story to the class.

Why does the raccoon have a bandit's eyes or a striped tail? Why is the hummingbird's throat red? Choose one of your favorite North American animals and make up your own story about its most unusual characteristic.

Each region of the country has a different environment. Where would you like to live? What plants and animals would live there with you? What would you eat? Out of what materials would you make your home or clothing? What would you do for fun? What problems might you face? Imagine that you and your family lived here before Christopher Columbus came to Turtle Island. Write a story about an exciting or troublesome day. Build this into a story as if it really happened. Revise, rewrite, edit, and type your story to share with your class.

From *Learning from the Land: Teaching Ecology through Stories and Activities,* Second Edition by Brian "Fox" Ellis. Santa Barbara, CA: Libraries Unlimited. Copyright © 2012.

Note

1. In an August 2010 issue of *Time* magazine there was an article on "Animal Intelligence: Birds that Use Tools," with a video of a rook placing stones in a pitcher! *Time* magazine archives, http://www.time.com/time/video/player/0,32068,490943046001_2009877,00.html.

Bibliography for Further Research

Casanova, Mary, illus. by Ed Young. *The Hunter*. Atheneum Books for Young Readers, an Imprint of Simon & Schuster, 2000. 0-689-82906-X. A young hunter saves a snake from the beak of a crane and is given the gift of tongues, the ability to understand animal languages. He uses and loses this gift when he saves his village from natural disaster.

Casler, Leigh, illus. by Shonto Begay. *The Boy Who Dreamed of an Acorn*. Philomel Books, 1994. 0-399-22547-l. This is a beautiful story that is easily adapted for the telling, about a boy's vision quest and his search for identity. The oak is more than a metaphor, a real teacher about wisdom and patience and providing for your community.

Cherry, Lynne. *The Great Kapok Tree*. Harcourt Brace Jovanovich, 1990. 13: 978-0152026141. The creatures of the rain forest share their wisdom with the woodcutter who is there to fell the mighty Kapok Tree. This story is easy to act out as a skit with the entire class having roles as various animals.

Patterson, Katherine, illus. by Leo and Diane Dillon. *The Tale of the Mandarin Ducks*. Lodestar Books, 1990. 0-525-67283-4. A beautiful story of a young couple who defy the greedy lord to help a pair of mandarin ducks and find that their kindness is repaid in a surprising manner!

A Night on the Hunt: Wildlife Stories

It is the story of all life that is holy and is good to tell.

Black Elk, from *Black Elk Speaks*

Comments to the Teacher

THIS STORY IS ACTUALLY TWO vignettes that explore predator–prey relationships and backyard wildlife. The theme underlying both stories is that you do not have to travel to the Amazon jungle or the African savanna to find magnificent wildlife. Even in the most urban of environments you can find exciting and scary creatures that have not been fully understood or studied.

"A Night on the Hunt" is presented as one story. You could easily tell it that way, or tell one vignette at the beginning of a unit and tell the other vignette later in the unit to give students a second example of wildlife in their own backyard.

You may notice that the part of the story about the owl is similar to the story of the owl in "The Ballad of Rusty and Nancy." That story gives an overview; this story focuses on one episode to make a different point.

The owl vignette in this story can be used as an example of a story in which pictographs are used as a writing strategy to teach science research skills. (For more information on this aspect of the story, see page 5.)

The vignette about the spider is based on both field research and library/Internet research. The activity at the end of this story encourages both.

Like some other stories, "A Night on the Hunt" mentions some gory details about wildlife survival. Please preread the story carefully to be sure it is appropriate for your students. Remember, you are free to omit, downplay, or *play up any* details or scenes that you think will enhance your lesson and your students' enjoyment of the story.

And now the story . . .

A Night on the Hunt

The great horned owl is one of the most efficient hunters in America. This type of owl is found throughout the Americas, from northern Canada to Tierra del Fuego at the tip of South America.

The great horned owl lives in forests, prairies, mountains, and cities. Its wings are more than five feet long—longer than your arms. You might think that an owl with such long wings would have trouble maneuvering in the forest. Not at all. In spite of its long, long wings, the great horned owl soars through the trees. Dodging left and right, stretching its wings straight up and down, it moves among the huge trees in the forest.

A great horned owl has special feathers on the front of its wings that slice the air. You may have heard other kinds of birds flap their wings, but you will never hear a great horned owl flap its wings. Those special feathers on the front of its wings act like silencers, so it can fly without making a sound. It can sneak up on its prey, flying silently through the forest at night.

Great horned owls fly on silent wings under the moon, hunting for mice, for snakes, for squirrels, for skunks. Yes, they eat skunks, because owls do not have a sense of smell. If a skunk sprays the owl, it does not notice. It's all meat to him. But its favorite food is mice. On an average night, a great horned owl may eat 15 to 30 mice.

The great horned owl sings out, "Whoo-oo-whoo, whoo, whoo, whoo, whoo." Have you tried the call of the great horned owl? Listen first: it's one-two-three, one-two, one-two: "Whoo-oo-whoo, whoo-whoo, whoo-whoo." Now you try it. "Whoo-oo-whoo, whoo-whoo, whoo-whoo."

Great horned owls have big eyes so they can see in the dark. But sight is not their most important sense. Primarily, they use their ears. Way off in the distance, the great horned owl hears this sound: squeak, squeak. It hears a mouse squeaking more than 100 yards away. That's how great horned owls locate their prey: by using their ears. Then they fly in close. With those large eyes they see in the dark. The owl's ears and eyes are an incredible pair of tools to use in the hunt. And let's not forget those talons. The great horned owl's talons are sharp as needles and as powerful as claws.

Far away, the owl hears this sound: squeak, squeak. It hears a mouse. The owl jumps from its branch, stretches its long wings, and begins flying silently through the forest.

For just a second, close your eyes and imagine you are a mouse. Imagine you have little paws, a little tail, and little whiskers. Open your eyes. The little mouse is eating little seeds, little bits of . . . *snatch!* From out of nowhere the owl swoops down. With its razor-sharp

talons it picks up that mouse. Guess what that owl is going to have for supper?

You might think that's pretty sad for the mouse, but think about it another way. Maybe it's not so sad. You see, that mouse became part of the owl. The owl could fly through the forest with the strength it got from eating the mouse. Because of the mouse the owl could live.

The truth is, if you are looking for adventure, you don't have to go to Siberia to study tigers, and you don't have to go to Australia to study the great white shark. There is plenty of adventure and excitement in your own neighborhood.

Most people do not walk around their neighborhood at four o'clock in the morning. If they did, they might hear owls calling. If they learned the call of the owl, they might be able to call one in closer so they could see it. Owls live in the parks and wild lands in and around the city. They hunt in your neighborhood.

Wildlife adventures are happening in your part of the world, wherever you live.

Let me tell you another adventure, one that happened right inside my house.

I was sitting at my computer in my basement working on this story about an owl. As I was working, I heard this strange sound: tick, tick. I wondered what was making that sound. I thought at first that maybe it was the clock making that ticking sound. I have an old cuckoo clock I bought one time when I was telling stories in Germany and it makes funny sounds sometimes.

I got up from my desk to explore the noise. I walked toward the sound. It wasn't coming from the clock, it

was coming from under a table. Using my ears, I was able to follow the sound and find the source of this strange noise. I looked underneath the table. The ticking sound was coming from there.

I got down on my hands and knees to take a better look. It was dark under there, so I ran and grabbed a flashlight. I came back and put the spotlight on a huge spider. It was a wolf spider—one of the big, hairy ones. This wolf spider had wrapped a web around an insect. I could tell some kind of beetle was trapped in that web, but it was so entirely covered with the web that I could not tell what kind of beetle it was. But I could tell that beetle was making the ticking noise.

Here's what was happening: the wolf spider would wrap the beetle up in the web, but the beetle would break the web. That was the ticking sound I heard. It was sort of like a cricket rubbing its legs together. The wolf spider would wrap up the beetle and *tick*, the beetle would break the web. The wolf spider would wrap it up and *tick*, the beetle would break the web. The wolf spider would wrap it up and *tick*, the beetle would break the web. This just kept going on and on.

Finally, the spider wised up and stopped the silly game. It bit the beetle and injected poison into it. Then

it wrapped up the beetle in its web again. *Tick*, the beetle broke the web. The spider wrapped it up and *tick*, the beetle broke the web again. The spider wrapped it up and *tick . . .*

Again the spider bit the beetle. Again it wrapped up the beetle. *Tick . . .* wrapped it up and *. . . tick*. The ticks came slower and slower. The spider bit the beetle again, and the ticks became weaker. Finally, the beetle couldn't break the web.

The spider wrapped up the beetle so well you could not see it at all. It was like a mummy in a cocoon, it was so thoroughly wrapped. Then the spider bit it one last time. You could see the spider's fangs; they were huge compared to the beetle.

I was so intrigued by this adventure that I went to the library and found a couple of books on spiders. I began to read about spiders and their poisons. Different types of spiders have different types of poisons. Some spiders have two kinds of poison. The first poison kills the victim's nerves. Nerves are like a wiring system in your body. When you want to move your finger your brain sends an electrical or chemical pulse, both at different junctures. It goes from your brain down your spinal cord, down your arm, into your hand, and makes your finger move. Nerves make up the wiring system that tells the muscle to move. When you kill an insect's nerves, it becomes paralyzed.

The other kind of poison is like an acid, and it dissolves the insides of the bug. All of the insides of the bug become a sticky goo, so the spider can come back later, insert its fangs, and like a straw, *sluuurrrrp*, suck it up, like a bug milkshake!

Next, I began to do a little scientific study. I went around my house looking for spiders and insects. I know and you know that no matter how clean your house is, insects move into all available spaces. Actually, spiders are great bugs to have in your house, because if you have spiders around, you do not have as many mosquitoes, or other kinds of bugs. The spiders keep them in check. If you have lots of spiders, if you find spiderwebs in your house, that's a good thing. You may not think so, and your parents might not think so, but believe me, it is.

I began to go around my house and look for different kinds of spiders. I found some that make webs to trap their prey and others that stalk, like tigers, and pounce upon their victims. Where there is water, like a lake or river, you will find lots of bugs, and where there are lots of bugs, you will find lots of predators. I found lots of spiders. I found 15 different species of spiders in and around my house.

One morning, I went out on my back porch and saw a web that was three feet across. It was the web of a garden spider. This kind of spider grows to be three inches long, with a beautiful pattern of yellow, black, and white on its back.

Inside my house I found little jumping spiders. Have you seen those? They are black and white, and they hobble around like little crabs. Garden spiders make webs to catch their prey, but jumping spiders are predators more like tigers; jumping spiders stalk their prey. They move very slowly, and when they get close to their prey, *they pounce on it!* They catch it. They bite it. They paralyze it. And you know what happens next. They inject their prey with acid to dissolve its insides and turn it into a milkshake. *Sluuurrrp!*

Several times I looked under the chair in my office and found additional mummified insects. After I was sure the spider had moved on to a new location, because I did not want to disturb it, I plucked several of the digested bugs from its web. I carefully pulled apart the cocoons. I had some trouble identifying some of the insects. There were three houseflies and two small wood roaches that I could identify for sure.

The wolf spider did me a big favor by catching those annoying bugs. And it also taught me something important: that I don't have to travel halfway around the planet to see wondrous wildlife. Exciting adventures are happening right inside my home!

Epilogue: Predators hunt in our neighborhoods every day. If you were to go out at sunset or sunrise, you might find owls hunting. But you don't have to go outside to find hunters: some predators and their prey are living right in your house. Have you spent time watching the spider behind your couch? Have you seen foxes or raccoons in your neighborhood? Coyotes? Hawks or owls? Have you seen robins hunt for worms? Did you know they use their feet to feel the vibrations a worm makes as it moves underground? Over the next few days or weeks keep your eyes and ears open. Watch for wildlife that has adapted to the city. You might find more excitement than you think!

Follow-Up Ideas for "A Night on the Hunt"

Discussion Topics

Ask students if they would like to be a wildlife biologist who studies wild animals. Tell them you will give them a chance in just a few minutes, but first, let's talk about some of the wildlife we might find in our neighborhood. Have they seen lions or tigers or bears, oh my? Have they seen smaller predators? Where? What might we see and what could we learn from these animals?

Writing and Telling the Story of Wildlife

❖**Grade Levels**: K–12   **Time estimate**: 50–60 minutes

❖**Science skills**: Communication; Observation; Inference; Reorder, Analyze, and Draw Conclusions

❖**Objectives**: Students will refine observation skills and exercise inferencing abilities based on these observations.

Students will model creative nonfiction based on field ecology, and the study of wild animals in the wild via live webcams.

Their field notes will be translated into dynamic stories about the lives of their fellow creatures.

National Standards

Science Standards

NAS 1 Science as Inquiry: Abilities necessary to do scientific inquiry; Understandings about scientific inquiry.

NAS 3 Life Science: Structure and function in living systems; Reproduction and heredity; Regulation and behavior; Populations and ecosystems; Diversity and adaptations of organisms.

Language Arts Standards

NCTE 1 Students read a wide range of print and nonprint texts to build an understanding of texts, of themselves, and of the cultures of the United States and the world; to acquire new information; to respond to the needs and demands of society and the workplace; and for personal fulfillment. Among these texts are fiction and nonfiction, classic and contemporary works.

NCTE 4 Students adjust their use of spoken, written, and visual language (e.g., conventions, style, vocabulary) to communicate effectively with a variety of audiences and for different purposes.

NCTE 5 Students employ a wide range of strategies as they write and use different writing process elements appropriately to communicate with different audiences for a variety of purposes.

NCTE 7 Students conduct research on issues and interests by generating ideas and questions, and by posing problems. They gather, evaluate, and synthesize data from a variety of sources (e.g., print and nonprint texts, artifacts, people) to communicate their discoveries in ways that suit their purpose and audience.

NCTE 8 Students use a variety of technological and information resources (e.g., libraries, databases, computer networks, video) to gather and synthesize information and to create and communicate knowledge.

NCTE 11 Students participate as knowledgeable, reflective, creative, and critical members of a variety of literacy communities.

❖**Materials**: Pen or pencil and paper; nature video that focuses on one animal species; DVD and television; and a copy of *All Upon a Sidewalk* by Jean Craighead George

Instructional Procedures

Introduction: To help students learn wildlife research skills and to improve their creative writing skills, students observe an animal in a video, then use what they have learned to write a story. Begin with an oral reading of *All Upon a Sidewalk* by Jean Craighead George or a story that you have written about a day in the life of a ferocious beast. Preface this reading with the idea that this story is entirely true, nonfiction, but it uses creative ideas to see the world as an ant might.

Activity: Tell the class that you are all going to write a story together. As a class, watch a National Geographic or Animal Planet DVD that focuses on one animal. Ask the class to take notes about the animal's physical characteristics; habitat; food supply; predators; defenses; mode of locomotion; and unique characteristics, such as molting, migration, or hibernation. Every five minutes or so, stop the video so that you and the students can take notes. You may want to model note taking during the first break by writing your notes on the board.

After watching the DVD, guide the class as it constructs a story about the animal. Call the story "A Day in the Life of . . ." Help students translate facts and ideas from their notes into poetic sentences with descriptive words and action verbs. With older students discuss the differences between first-person narration and omniscient narration. Help students describe and develop a clear and specific setting. Introduce ideas like character development, suspense, conflict, climax, and resolution. Encourage students to imagine being the creature, to present the facts from its point of view, and to speak in its voice. Working as a whole class, create a rough draft on the smart board, chalkboard, or overhead.

As a class, review the story, looking for ways to add color, suspense, or important details. Encourage students to review their notes and offer suggestions for details to be added to the story. Cut out or tone down flowery sentences and rearrange paragraphs, modeling the art of rewriting, refining, and editing. Double-check spelling, punctuation, subject-verb agreement, and other grammatical details.

Finally, tell the students their story! When you add dramatic interpretation and sound effects, the vitality of an oral presentation will make students beam with the success of a story well written.

For homework, send students to any one of a dozen live webcams of wild animals online. From owls and eagles, to zoo pandas and elephants, there is a wide world of wildlife just a few clicks away. Admittedly, these do not work so well for a classroom lesson, as they can be frustrating when there is nothing there. But many live webcams also archive some of their best footage, which can be previewed and saved for classroom use. In the same manner as the classroom exercise, students are to spend some time observing the animal and taking notes. Their observations are the raw material for their rough draft. They can do additional research online and turn their ideas into a polished story with a little re-writing and editing. National Geographic: http://animals.nationalgeographic.com/animals is a great place to start.

Assessment: Students stories can be evaluated based on their creative interpretation of the scientific material as well as writing mechanics.

Follow-Up Activities: Virtual reality is never as good as reality! Take your students outdoors to research real animals!

Wild Animal Observation

❖**Grade Levels**: K–12

Time estimate: Several days, a few minutes a day

❖**Science skills**: Communication; Observation; Prediction; Inference; Identify Variables; Formulate Hypotheses; Reorder, Analyze, and Draw Conclusions; Design Investigations

❖**Objectives**: Students will develop observation skills as they conduct research on backyard wildlife.

Students will learn to write creative nonfiction that explores the lives of wild neighbors.

National Standards

Science Standards

NAS 1 Science as Inquiry: Abilities necessary to do scientific inquiry; Understandings about scientific inquiry.

NAS 3 Life Science: Structure and function in living systems; Reproduction and heredity; Regulation and behavior; Populations and ecosystems; Diversity and adaptations of organisms.

Language Arts Standards

NCTE 1 Students read a wide range of print and nonprint texts to build an understanding of texts, of themselves, and of the cultures of the United States and the world; to acquire new information; to respond to the needs and demands of society and the workplace; and for personal fulfillment. Among these texts are fiction and nonfiction, classic and contemporary works.

NCTE 4 Students adjust their use of spoken, written, and visual language (e.g., conventions, style, vocabulary) to communicate effectively with a variety of audiences and for different purposes.

NCTE 5 Students employ a wide range of strategies as they write and use different writing process elements appropriately to communicate with different audiences for a variety of purposes.

NCTE 7 Students conduct research on issues and interests by generating ideas and questions, and by posing problems. They gather, evaluate, and synthesize data from a variety of sources (e.g., print and nonprint texts, artifacts, people) to communicate their discoveries in ways that suit their purpose and audience.

NCTE 8 Students use a variety of technological and information resources (e.g., libraries, databases, computer networks, video) to gather and synthesize information and to create and communicate knowledge.

❖**Materials**: Wild Adventures in Your Backyard worksheet (pages 156), pen or pencil and paper

Instructional Procedures

Introduction: If the lessons in this unit have been followed in order then your students have had the training they need to be wildlife researchers!

Activity: To give students field experience as wildlife researchers, instruct them to observe a *wild* animal in their neighborhood. The wild animal could be an ant, a squirrel, a bird at the feeder, or a spider in the basement. (Pets are not wild animals.) Challenge students to take careful notes with a time log, as if they were wildlife researchers. Tell students to focus on the animal's physical characteristics; habitat; food supply; predators; defenses; mode of locomotion; and unique characteristics, such as molting, migration, or hibernation.

After students complete their research, coach them through the process of writing stories about the animal. Have students illustrate their stories and make them into books. Early elementary classes can write their stories using pictographs (see below).

Students who dislike writing in any form may warm to this project. In a similar project, one student who hated to write—he even refused to write his full name on his math papers—filled several pages with a story about a tiger on the hunt! (His research was a combination of zoo visits and a DVD.)

Assessment: Collect their research notes to be evaluated based on the details of their observations. Their stories can be graded with the standard rubric on both science and language arts.

Follow-Up Activities: Working together as a class, you could publish a guide to local wildlife, create an art exhibit, or turn their stories into poems, (see below).

Say It with Pictographs

❖**Grade Levels**: K–3 **Time estimate**: 45 minutes

❖**Science skill**: Communication

❖**Objectives**: Students will develop preliteracy skills as they translate their scientific ideas into pictographic stories.

National Standards

Science Standards

NAS 1 Science as Inquiry: Abilities necessary to do scientific inquiry; Understandings about scientific inquiry.

NAS 3 Life Science: Structure and function in living systems; Reproduction and heredity; Regulation and behavior; Populations and ecosystems; Diversity and adaptations of organisms.

Language Arts Standards

NCTE 1 Students read a wide range of print and nonprint texts to build an understanding of texts, of themselves, and of the cultures of the United States and the world; to acquire new information; to respond to the needs and demands of society and the workplace; and for personal fulfillment. Among these texts are fiction and nonfiction, classic, and contemporary works.

NCTE 4 Students adjust their use of spoken, written, and visual language (e.g., conventions, style, vocabulary) to communicate effectively with a variety of audiences and for different purposes.

NCTE 5 Students employ a wide range of strategies as they write and use different writing process elements appropriately to communicate with different audiences for a variety of purposes.

❖**Materials**: Pen or pencil and paper

Instructional Procedures

Introduction: Writing can be difficult for those who do not have a firm grasp of phonetic language. After they have found a symbolic representation for their ideas (in the form of pictographs) and have communicated their story verbally, writing becomes easier.

Activity: Ask students to create their story first as a series of sequential pictures. This helps them to create well-ordered stories that follow a natural progression from beginning to end. It also helps them to find words to use in telling their story.

With the pictographs completed and arranged in the proper order, ask the students to say aloud what is happening in each picture. This will help them bridge the gap between the spoken and written word.

Assessment: An oral presentation of their stories is the best way to evaluate their language skills in this lesson.

Follow-up Activities: Finally, ask the students to put their stories into words on paper. For more information about pictographs, please visit this Web page:

http://www.inquiry.net/outdoor/native/sign/pictographs.htm

(The author's pictograph for "A Night on the Hunt" appears below.)

Guessing-Game Poems

❖**Grade Levels**: 1–12   **Time estimate**: 60 minutes

❖**Science skills**: Communication; Classification; Inference; Identify Variables

❖**Objectives**: Students will demonstrate deductive logic skills as they write and guess at riddles.

National Standards

Science Standards

NAS 1 Science as Inquiry: Abilities necessary to do scientific inquiry; Understandings about scientific inquiry.

NAS 3 Life Science: Structure and function in living systems; Reproduction and heredity; Regulation and behavior; Populations and ecosystems; Diversity and adaptations of organisms.

Language Arts Standards

NCTE 1 Students read a wide range of print and nonprint texts to build an understanding of texts, of themselves, and of the cultures of the United States and the world; to acquire new information; to respond to the needs and demands of society and the workplace; and for personal fulfillment. Among these texts are fiction and nonfiction, classic, and contemporary works.

NCTE 4 Students adjust their use of spoken, written, and visual language (e.g., conventions, style, vocabulary) to communicate effectively with a variety of audiences and for different purposes.

NCTE 5 Students employ a wide range of strategies as they write and use different writing process elements appropriately to communicate with different audiences for a variety of purposes.

NCTE 7 Students conduct research on issues and interests by generating ideas and questions, and by posing problems. They gather, evaluate, and synthesize data from a variety of sources (e.g., print and nonprint texts, artifacts, people) to communicate their discoveries in ways that suit their purpose and audience.

NCTE 8 Students use a variety of technological and information resources (e.g., libraries, databases, computer networks, video) to gather and synthesize information and to create and communicate knowledge.

❖**Materials**: Pen or pencil and paper; research materials, including encyclopedias of animal behavior, the Internet, DVDs, and CDs

Instructional Procedures

Introduction: Challenge students to exercise higher-level thinking skills and research animal behavior by writing poems that describe the behavior of a particular animal but do not name the creature.

Activity: Ask students to choose an animal, preferably one that lives in their ecosystem, possibly the one they researched in one of the other lessons in this chapter. Ask them to research the animal online, in the library, or via DVD, and then make a list of its behaviors, that is the things the animal does, thinking in terms of verbs. Students use this list of words to write a brief poem that describes the animal's behavior. Remind students to avoid naming the animal in the poem.

Students can read their poems to one another and let their listeners guess what animal is being described.

Assessment: The trick here for the students and for the teacher evaluating the poem is this: Did they give good clues without giving it away too soon?

Follow-Up Activities: Another way to use these poems is to create a guessing-game bulletin board. Have the students fold a piece of paper in half. Tell them to write their poem on the front of the paper and draw a picture of the animal on the inside. Hang the sheets on the bulletin board so that students can read the poems. Students try to guess what animal is being described, and then peek at the illustration to verify their guess. They can also be typed, edited, and bound into an animal guessing-game booklet with desktop publishing.

Following is an example of a guessing-game poem:

A Midnight Visitor

Bandit creeping through the night,

Click, click, click, I flick the light.

She scurries off, climbs up the tree,

Near the roof she looks down at me.

Her eyes reflect the flashlight beam,

She hisses, shrieks, her teeth they gleam.

I close the door, turn off the light.

She scurries off into the night.

Wild Adventures in Your Backyard— YOU Can Be A Wildlife Researcher!

In your neighborhood, maybe even in your home, are hundreds of wild animals living amazing, dangerous, and curious lives that we know very little about. When you carefully observe these animals, they will reveal to you their habits and adventures. Animal, habitat, and behavior are the elements of any good research project. The animal you choose is the character, the habitat is the setting, and the animal's behavior is the plot. Do some field research. Study the wildlife in your home or neighborhood and turn its actions into an exciting story. Use this sheet to organize your research. Write your notes in a journal or on the back of this sheet of paper.

Be sure to choose an animal that you can easily observe almost any day. A bird at your feeder, an ant on the sidewalk, or a squirrel in a park nearby; these are just a few possibilities.

My animal is a _____

Character: Observe your animal for a few days. Follow it if you need to, but don't interfere with it. Write a detailed description of what your animal looks like, how it moves, what it does, and any other interesting characteristics.

_____
_____
_____

Setting: The second step in your research is to spend some time in the animal's environment or *habitat*. Jot down some notes about its home. What types of plants and animals share its neighborhood? Is there a creek or river or any unusual rock or land forms nearby? Write a detailed description of your animal's habitat or home.

_____
_____
_____

Plot: Schedule regular times to sit quietly and unobtrusively in the animal's habitat. Watch the animal very carefully. Ask yourself questions like: What does it eat? What might eat it? Where will it hide? Where does it sleep? What does it do in bad weather? Does it have babies? What materials does it use to build its home? Make up some questions of your own and then observe your animal for answers. The basic question is: How does this animal interact with its environment? Note everything that your animal does, and then make a prediction or hypothesis about what you think it will do next. Watch carefully to see if your guess was accurate. Observe your animal at least 15 minutes a day for seven days. Be careful not to disturb its daily habits. Always take careful notes, recording the time, the date, the weather, and any other pertinent information. You may also wish to supplement your observations with information from the library or Internet.

_____
_____
_____

The Story: Use your notes to write an exciting story about your animal. You may wish to begin with a description of the setting or character, plunge into a problem the animal faces, build tension, and then tell how the animal resolves the problem.

From *Learning from the Land: Teaching Ecology through Stories and Activities,* Second Edition by Brian "Fox" Ellis. Santa Barbara, CA: Libraries Unlimited. Copyright © 2012.

Bibliography for Further Research

Baker, Nick. *Nature Explorers: Backyards and Parks*. Harper Collins, 2007. 978-06-089079-7. In a fun, user-friendly format, this book provides you with skills and tools to be a backyard researcher!

Paulos, Martha. *InsectAsides: Great Poets on Man's Pest Friends*. Viking, 1994. 0-670-85567-7. From ee cummings to Pablo Neruda, this is a great collection of truly spectacular poetry.

Ruddell, Deborah. *Today at the Bluebird Café*. McElderry, 2007. 13:978-0689871535. Funny and playful, Ruddell has put together a clever collection of poetry about birds.

Sidman, Joyce, illus. by Beth Krommes. *Butterfly Eyes and Other Secrets of the Meadow*. Houghton Mifflin, 2006. 13-978-0-618-56313-5. Gorgeous illustrations bring to life several well-crafted riddle poems, each followed by a brief, fact-filled essay about the creatures alluded to in the poems. A good model for student poetry and essays.

There is a growing collection of well-written books about the daily lives of animals that fall into the categories of creative nonfiction or realistic fiction. Here are a few of my favorites:

Arnosky, Jim. *Deer At the Brook*. Mulberry Press, 1986. 13: 978-0688104887.

Arnosky, Jim. *Raccoons and Ripe Corn*. Mulberry Press, 1987. 13: 978-0688104894.

Bonners, Susan. *A Penguin Year*. Dell, 1981. 13: 978-0440401513.

Freschet, Berniece. *Bear Mouse*. Charles Scribner's Sons, 1973. 13: 978-0437413031.

Winter, Jeanette. *The Tale of Pale Male*. Harcourt Books, 2007, 978-0-15-205972-9.

Yolen, Jane. *Owl Moon*. Philomel Books, 1987. 13: 978-0399214578.

The Salmon and the Stream: A Story of Animal Migration by Garth Gilchrist

We heed no instincts but our own.

—Jean de la Fontaine, from *Fables*

Comments to the Teacher

THIS IS THE ONLY STORY in this collection that I did not have a hand in writing. I use it first and foremost because it is an excellent piece of writing and beautifully expresses many of the same goals that I strive for in my stories.

Garth and I met at an environmental education conference a few dozen years ago. We were on parallel paths, struggling to give voice to the creatures of the earth and sky, to evolve past the Romantic idea of using nature as a metaphor for human feelings. We were seeking a way to allow the authentic voice of nature to share its wisdom. Our goal was to allow our writing to move away from animals as human-like characters into a realm where the wisdom of other creatures is respected. We discussed ways to bring science theory to life and to ground environmental activism in hard science. I shared with him my story "The Web." Together we discussed "Walter the Water Molecule." Several years later he sent me a copy of "The Salmon and the Stream."

With his permission, I share his story about the migration of the Pacific salmon.

Although this story is anthropomorphic in that it gives the salmon human-like feelings, thoughts, and dreams, the author's goal, which he beautifully achieves, is to see the world through the salmon's eyes. Because we do not fully understand the interior processes, thoughts, and communications of animals, we are limited in how we can write about them. Garth uses a sophisticated narrator's voice, an omniscient internal consciousness, to express the mystery of instinct. He assigns human feelings and thoughts in an effort to show possible reasons for the salmon's behavior.

When I tell this story I am reminded of the traditional Nootka tale of the salmon boy who falls into the river and becomes a salmon. After living with his relatives under the sea, he comes back to the river to be caught by a fisherman from his village. The village elder helps the family to return their son to his human form. Now a man, he has a deeper understanding of the salmon's life. In both the traditional tale and the following story there is a profound respect for all beings.

And now the story . . .

The Salmon and the Stream

At the edge of a mountain stream, tiny waves of water lapped against a sandy bank. Out in the stream, under the clear water, and beneath the sand a few inches down, a hundred tiny red eggs lay together in their dark nest. The sand had protected them all winter from the rushing current, and from the snow that covered the stream. Now, as the last snow melted, an egg was stirring.

Inside the egg, a tiny salmon sensed that life was beginning. As the water warmed, the half-inch long salmon thrashed inside its slippery egg prison, until finally, the walls that trapped it burst open. The newborn fought her way up through the sand. Soon she found herself swimming free in the clear water. A huge cutthroat trout lunged toward her, its mouth wide open and hungry, but instantly the miniature tail flicked and the young salmon disappeared behind a huge gnarled root growing in the bank. She waited there until she sensed the danger had passed. Then she swam back out into the pool. Before long several other tiny fish shared the water. Then there were dozens, then hundreds of silvery flashes in the water, all just hatched, tiny slivers of light, darting this way and that in the streamside that was to be their home for the next few months.

The newborn salmon drew water continually into her mouth and passed it over her gills, drawing the life-giving oxygen out of the water and into her body. And she ate the food the stream brought her, bits of drifting insects and tiny floating plants. The river gave her life and she grew. During the days that followed she stayed in the shallows, in the calm shimmering pools along the edge of the swift-flowing stream, or in the tall grasses and thick reeds that grew in the water along the banks. When large fish appeared, the salmon hid behind the huge gnarled root. Out in the main current of the stream, the deep water swelled and tumbled. Inside the young fish, a quiet voice said, "Stay back, stay back." The salmon obeyed the voice.

As the days passed the hatchling continued to feed, and she grew. Her body was streamlined and sleek. She could dart quickly even upstream against the current. Her tail and fins she moved precisely, guiding herself to one side or the other of reeds and roots or between the shining sides of the other young fish that shared the pool with her.

One day, the voice inside the salmon said, "Swim into the current, swim into the current." She hesitated, for she had never left the quiet pool. But the voice urged her, "Swim out, swim out," so she swam, out into the current, and she let it carry her strong body down the rushing stream. As she swam, the stream tumbled her down rapids and threw her over waterfalls, but her tough salmon skin and silvery scales slid easily over the rocks. She allowed herself to be carried by the water. She ignored the deep green pools, and the eddies of big polished boulders that might have given her safety and rest. Inside her, the voice kept chanting, "Swim with the current, swim . . ."

The quiet voice was always talking. The quiet voice spoke not in words but in feelings and in pictures that drifted through her mind. She always obeyed the voice. It was a part of her.

Sometimes, she dreamed. In her dream she was floating in a vast, deep pool—a pool with no bottom and no banks. There was only water, water, water everywhere, endless blue-green high above, far below, and all around.

The stream grew wider now as other streams flowed into it, swelling it with fresh currents and new smells. It roared over the rocks and swept powerfully around the bends, and one day the rushing stream plunged over a small waterfall into a wide river. As the young salmon felt herself falling and then felt the new, deep rolling waters about her, excitement welled up within her.

At the beginning of her journey, the young salmon had heard only the trickle of the shallow, running water over the little pebbles and stones, and the whoosh of water through roots and reeds. Then as the stream swelled she'd heard the roar of rapids. Now she heard the silence of the deep flowing river, washing smoothly but powerfully over the round stones along the river bottom.

A few days passed and a new sound filled the water. The sound started as a whisper in the silence, but the strange sound grew louder and louder. It was like the roaring of a rapids but it throbbed and pulsed. The roaring grew until it filled the water around her like a huge heart, pounding, pounding, louder, crashing. "What is it?" thought the salmon. Fear flashed through her but the voice said, "Go on."

All at once the roaring was all around. The water heaved and tumbled. It crashed and swelled. The water was dark, filled with sand. The water was thick and tasted of salt. "Swim out, swim out!" shouted the voice. And the salmon swam out, through the breaking waves.

Slowly, slowly, like a dream, the water cleared, became deep and blue-green. The banks of the river vanished. The current disappeared. The water was still, but vast. The young salmon could see no bottom. There was only water, water everywhere. And then she remembered the dream! Yes, this was like the dream. This place was endless, quiet and shining like the dream had been. She felt at home here in this place. It was new, but somehow she knew it.

In the days that followed the salmon forgot the current, and she forgot the river. She gave herself to the ocean. The voice was quiet. "Yes," it whispered, "yes."

There were whales here in the ocean. Yes, the wide sea was her new home now, and there were whiskered seals, and fish of many colors and strange creatures. The young salmon was alert. She sensed danger, yet it seemed right that all these things were here, a part of this place.

Other salmon had tumbled down the stream, too—many others! The quiet voice within the young salmon said, "Swim with the others, stay close to them." She obeyed the voice and swam with them. Together, they were like a shimmering cloud, shifting shape and direction in the blue-green sea sky. They were always together. Moving with the others, the young salmon sought food, or just drifted. Seals would dart in and grab other salmon right beside her with their glistening white teeth, but she darted quickly and was never caught.

The light came each day, each morning, and each evening the darkness; day to night, and night to day, day to night, and night to day

Sometimes the salmon flowed with the ocean currents, great rivers of the sea, following the food: smaller fish, shrimp, and jellyfish. The salmon let the current carry them for weeks until the quiet voice said, "No farther, swim back now." They never questioned the quiet voice. It was a part of them. It told them how to be salmon.

Months passed. A year passed. Two years. All at once the young salmon knew: something important was about to happen.

The salmon was large now, and strong. In the last few months, eggs had begun to grow within her belly, swelling her size. Soon the eggs would want to be laid. "But not here. No, not here."

Once more the salmon began to dream. She dreamt of being hurled and tossed in the sandy waves. In her dream, she heard the crashing of the waves, the pounding on the shore. The voice inside her said, "Swim, swim to the waves."

The salmon swam to the east. Not quite east. A little to the north, too. The voice told her it was the right way. And there it was! She could hear it now, the thundering crash of the waves. She plunged into the sandy water. Thousands of salmon surrounded her, all heaving in the waves. "Yes!" said the voice. "Go on, go on!" The salmon swam on. A way opened through the sand. Now there was fresh water, sweet clear water. The salmon swam into it. She saw the bottom, the rocks, the banks. She smelled the sweet, flowing water. She remembered this, excitement bursting within her. "Yes," said the voice. "Push

now. Push against the water. You must swim against the water."

The salmon swam. She dreamed as she swam. She dreamed of the streamside where her life had started. She dreamed of still water at the edge of the flowing current. She dreamed of the sandy bottom, the reeds and the grasses in the sand, the place she had come from. And she felt the eggs within her. "Go there," said the voice. "Swim!" And with the dream in her mind, she swam.

Many streams flowed into the river. The river grew cooler and the salmon smelled strange, new, but familiar scents. At each new stream mouth salmon would leave the flowing river and enter the new water. But the voice inside her said, "No, stay in the river."

Her body began to change. First it grew pink, and as the days passed the pink deepened to red, as red as the eggs she had hatched from. But she didn't even notice. She only swam, pushing against the water. Hundreds of big, shining, flashing salmon swam around her, pushing their way upstream, sliding together over the rocks, through the clear water, dreaming of their beginning place, swimming against the current. Inside the salmon was her dream, a picture floating in her mind: the streamside, the still water, the sandy bottom, the reeds and grasses. "Find it," said the voice. "Find it!"

After many days, one morning there was a sudden smell that sent a shock through her whole body. "Follow it! Follow it," cried the voice. She leapt toward the scent of the new water. Water was pouring over a high rock. She leapt up, flashing through the air, and splashed into the clear pool above. It was her stream. The water was shallower, narrower. "Yes," said the voice, "yes, swim now, swim!"

Food passed by, but she did not notice it. She had not eaten for days. She didn't know it, but she would never eat again. "Swim," said the voice, "swim!"

And the salmon swam. The stream grew steep and fast. She beat her tail against the racing current. The water rushed by her. Thrusting, lashing her tail against the water, forcing her way upstream, between the rocks, under the logs, through the mazes of sticks. There were many fish around her. She paid no attention to them. "Swim!" insisted the voice. The rocks grew larger. The stream channel grew steeper still. The water foamed and sprayed and heaved. "Push!" said the voice! Push upward.

The stream roared around her. The water poured down over high, jagged rocks. The pool lay high above. Flashes of silver salmon shot into the air from the water like bullets. Many fell back. The water and the air were full of the shining bodies of salmon. She hesitated. "Jump up!" shouted the voice. She dove to the bottom, thrashed her tail, flew up through the water and shot into the air. Up she flew. She hit the rock. She fell back. "Again!" cried the voice. "Jump again!" Again she dove to the bottom, deeper now, thrashed furiously against the foaming water, sped through the water, broke the surface and exploded into the air.

Again she smashed into the rock, tumbled down it and slapped into the water. The eggs inside her were strong. "Again!" shrieked the voice. "Again! Again!" Once more she dove, circled the bottom, gathered speed, thrashing her tail with enormous force, sped upwards, streaked into the air, flew, higher, higher, higher this time. She splashed into the pool above the rock! "Yes," said the voice, calmer now. "Go on."

The water streamed down through steep gorges. The rains were falling, the stream swollen with new water. Again and again she leapt the rocks, leapt the logs, fought the raging water. Again and again she fell back, bruising and tearing her exhausted body. Her body was cut and she bled into the water. Her tail was torn, a fin was half ripped out, but she swam on. She swam because the voice said to swim.

Now the stream leveled, grew quieter, slower, and the salmon sensed an ending. The stream smelled like reeds, like grasses. "Watch," said the voice. "Watch carefully." The bottom grew sandy. The dream flooded her mind. The spot. The spot. Where is the spot? "No, not yet, not here," whispered the voice. "Swim. Swim." She swam slowly, and painfully. And then, she saw it. There was the root, the hiding place. "Yes! Yes!" cried the voice, triumphant.

Now a new dream flooded her mind. Another fish. A male. "There he is." said the voice. She looked and he was there, swimming near her. The female fanned the sandy bottom with her tail, hollowing a nest. She forced the shining red eggs from her body, the eggs she had carried within her to this spot for this moment. They settled into their sandy nest.

The male moved over the eggs now. His body quivered. The seed of life left him and drifted through the water over the eggs. They would grow. The male swam off and the salmon moved back over the eggs which she had just laid. She fanned the bottom, covering the eggs with sand.

Now she listened for the voice. "Rest," it said. "Drift." She let the water carry her. There was no dream. And now the voice was quiet. She drifted in the water. Her strength was gone. She could no longer swim. She let the water carry her. Waves washed the salmon up onto the beach. She lay there gasping for breath. "It is over," whispered the voice. "Rest now." An eagle came. The last of the salmon's strength, the eagle took for itself. The salmon's life had ended.

But not really.

The eggs she had laid were resting safely in the darkness under the water, beneath the sand. Bright, shining red eggs protected from the rushing current, and from the snow that fell in the months that followed.

Time passed, spring came, the snow melted. The ice broke, and the water grew warmer. Beneath the sand an egg began to stir. It sensed that life was about to begin. The tiny fish inside wiggled and lashed against the egg wall just as her mother had. The egg burst, and the newborn salmon struggled up through the sand into the clear water of the pool. The life of the salmon would continue.

Follow-Up Ideas for "The Salmon and the Stream"

Discussion Topic

Immediately following the story discuss the route of the Pacific salmon using a large map of North America. How do they know where to go? How do they find their way? Using the same map, discuss ways in which other animals migrate, including birds' journeys north and south, elk and bighorn sheep migrating into higher altitudes in the summer and downslope in the winter, and monarch butterflies' movements to Mexico.

Migration Route Map Study

❖ **Grade Levels**: K–12 **Time estimate**: 45–60 minutes

❖ **Science skills**: Classification; Communication; Prediction; Inference

❖ **Objectives**: Students will learn mapping skills.

They will demonstrate an ability to chart and measure their movements and empathize with the migrations of other animals.

National Standards

Science Standards

NAS 1 Science as Inquiry: Abilities necessary to do scientific inquiry; Understandings about scientific inquiry.

NAS 3 Life Science: Structure and function in living systems; Reproduction and heredity; Regulation and behavior; Populations and ecosystems; Diversity and adaptations of organisms.

NAS 5 Science and Technology: Abilities of technological design; Understanding about science and technology; Abilities to distinguish between natural objects and objects made by humans.

NAS 6 Science in Personal and Social Perspectives: Personal health; Populations, resources, and environments; Natural hazards; Risks and benefits; Science and technology in society.

Language Arts Standards

NCTE 4 Students adjust their use of spoken, written, and visual language (e.g., conventions, style, vocabulary) to communicate effectively with a variety of audiences and for different purposes.

NCTE 11 Students participate as knowledgeable, reflective, creative, and critical members of a variety of literacy communities.

❖**Materials**: Floor plan of the school; a road map of the neighborhood; pen or pencil and paper; a large map of North America and South America; embroidery thread in various colors; pushpins in various colors

Instructional Procedures

Introduction: Help students build a conceptual framework to understand this story by asking them to discuss their daily, weekly, or annual migration patterns. Remind students that migration means moving from one place to another with purpose, regularly and periodically. It does not mean that every time you go somewhere you are migrating. When students discuss their personal migration patterns, encourage them to differentiate between their movements that are migratory and those that are simply travel.

Activity: With a diagram of the school floor plan in hand or on the overhead projector, lead a discussion of the students' daily migration pattern—to lunch, to the restroom, outdoors for recess, and so forth. Ask students to discuss why they go to each area instead of doing everything in their homeroom.

Also ask students to discuss their weekly migration patterns to the gym, to art class, to the music room, and so forth. Again, ask for the reasons these trips are made. Throughout these discussions compare the students' migrations to animal migrations. For example, wildebeest migrate to the savanna during the rainy season, when the rains provide moisture for abundant grass; students migrate to the drinking fountain in the hallway because there is not a drinking fountain in the classroom.

With a detailed road map of the community in hand, ask students to chart their migration patterns for one day or one week. After the time period is up, ask the students to discuss the routes they took to school. Did they take the same route every time? Did they follow that route (or routes) regularly and periodically? Have students describe other trips they made and why those trips were made. Ask students to determine whether those trips were migratory or just travel. For example, a student visiting the allergist for a shot every week looks like migrating; a student who randomly went to see a friend one day simply traveled.

These discussions work best if students first show their map to a partner so everyone gets a chance to share with someone before a few students are asked to share with the entire class.

Again, these discussions could be laced with examples from the animal kingdom. For example, in the autumn a student goes to a department store to purchase a new winter coat. In the autumn caribou migrate to the boreal forest to escape the harsh winds that blow off the open tundra.

With a large map of North America and South America on the bulletin board, introduce the migration patterns of various animals. Using various colors of embroidery thread and pushpins, mark the routes of several species of birds, fish, and mammals. (A list of animals useful for this activity appears on the Animal Migrations worksheet, pages 00–00.)

Assessment: Students who participate in the conversation can be given bonus points for the following assignment.

Follow-Up Activities: These conversations are designed as a preview for the following lesson plan:

Writing a Migration Story

❖**Grade Levels**: 3–12   **Time estimate**: Two class periods of 45–60 minutes

❖**Science skills**: Observation; Communication; Inference; Design Investigations

❖**Objectives**: Students will research migration routes and demonstrate an understanding of how and why through their stories.

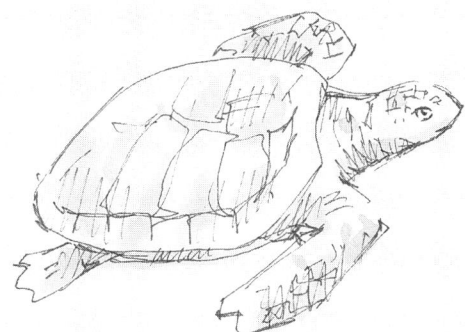

National Standards
Science Standards

NAS 1 Science as Inquiry: Abilities necessary to do scientific inquiry; Understandings about scientific inquiry.

NAS 3 Life Science: Structure and function in living systems; Reproduction and heredity; Regulation and behavior; Populations and ecosystems; Diversity and adaptations of organisms.

NAS 5 Science and Technology: Abilities of technological design; Understanding about science and technology; Abilities to distinguish between natural objects and objects made by humans.

NAS 7 History and Nature of Science: Science as a human endeavor; Nature of science; History of science.

Language Arts Standards

NCTE 1 Students read a wide range of print and nonprint texts to build an understanding of texts, of themselves, and of the cultures of the United States and the world; to acquire new information; to respond to the needs and demands of society and the workplace; and for personal fulfillment. Among these texts are fiction and nonfiction, classic and contemporary works.

NCTE 2 Students read a wide range of literature from many periods in many genres to build an understanding of the many dimensions (e.g., philosophical, ethical, aesthetic) of human experience.

NCTE 3 Students apply a wide range of strategies to comprehend, interpret, evaluate, and appreciate texts. They draw on their prior experience, their interactions with other readers and writers, their knowledge of word meaning and of other texts, their word identification strategies, and their understanding of textual features (e.g., sound–letter correspondence, sentence structure, context, graphics).

NCTE 4 Students adjust their use of spoken, written, and visual language (e.g., conventions, style, vocabulary) to communicate effectively with a variety of audiences and for different purposes.

NCTE 5 Students employ a wide range of strategies as they write and use different writing process elements appropriately to communicate with different audiences for a variety of purposes.

NCTE 6 Students apply knowledge of language structure, language conventions (e.g., spelling and punctuation), media techniques, figurative language, and genre to create, critique, and discuss print and nonprint texts.

NCTE 7 Students conduct research on issues and interests by generating ideas and questions, and by posing problems. They gather, evaluate, and synthesize data from a variety of sources (e.g., print and nonprint texts, artifacts, people) to communicate their discoveries in ways that suit their purpose and audience.

NCTE 8 Students use a variety of technological and information resources (e.g., libraries, databases, computer networks, video) to gather and synthesize information and to create and communicate knowledge.

❖**Materials**: Animal Migrations worksheet (pages 00); pen or pencil and paper; research materials

Instructional Procedures

Introduction: "The Salmon and the Stream" provides both scientific and empathic understanding of the life cycle of the Pacific salmon. One of the more important concepts introduced in this story is migration.

Activity: Writing in a similar style, students could write a story about another animal's life. Give each student an Animal Migrations worksheet. Allow each student to choose an animal to research and then develop a story.

Depending on their grade level and prior experience, students may need more or less assistance with Internet and library research skills, taking notes, and developing footnotes. There is an ever-growing number of science books for students, as well as some rather innovative interactive websites. This is one area that has grown exponentially since the first publication of this book. You can chart animal migrations in real time through Web pages like Journeys North. Encourage older students to explore the latest research through scientific journals available online or at the reference department of a local university library.

Exploring Point of View

Consider challenging students to experiment with point of view. In the introduction to the assignment, teach or review point of view.

Stories about animals work best in first person as the animal, or in third person, as an omniscient narrator. One of the strengths of Garth Gilchrist's story is that the omniscient narrator is aware of the interior voice of the salmon. This is a difficult voice to create, yet he masterfully maintains it.

At some point, as students are rewriting their stories, challenge the students to choose one lengthy paragraph and rewrite it using a different point of view. Compare and contrast the two points of view, including their strengths and weaknesses. For example, in a first-person point of view (i.e., in the animal's voice), it is harder to paint the big picture or provide scientific background without resorting to anthropomorphism. On the other hand, the omniscient, third-person narrator may miss some of the internal processes and feelings of the creature.

Anthropomorphism is one of the stickiest issues in writing about animals. Challenge students to stay within the realm of what science knows as fact while writing their stories. Animals are not humans, and we do not know exactly why they do things or how they think or feel. Give students room for some empathy and encourage them to attempt to see the world from the animal's perspective. The issue of anthropomorphism can be a very informative point of discussion and can provide serious exercises in rewriting. You may want to write the word on the chalkboard and discuss its meaning (i.e., anthropo = human, as in anthropology; morph = change, as in metamorphosis).

Emphasize to students that it is important to maintain a consistent point of view throughout the story. Be sure to coach them throughout the writing and rewriting process to be sure that they consistently use one point of view.

Assessment: The story could be graded both as a science report and as a language arts assignment. As a science report, the story must be accurate, with endnotes referencing the sources of information. Challenge them to measure the number of kilometers the animal travels. Credit students for describing the migration route in detail, not just A to Z, but also points B, C, D . . . W, X, and Y. Challenge students to explore the reasons why their chosen animal migrates, how it navigates from one point to another, any trials or tribulations the animal faces on its journey, and the effect of humans, both positive and negative, on this journey.

The language arts portion of the grade could emphasize the construction of an interesting story, with a dynamic beginning, climactic middle, and clear resolution. Consider giving students two language arts grades, one for the mechanics of spelling and grammar, and one for stylistic issues, such as sentence structure, transitions and flow, drama, and originality.

Follow-Up Activities: When students complete their stories, have them use heavy thread and pushpins to chart their animal's migration pattern on a map of the Americas (see previous activity). Ask students to tell their story to the class. Or, have students illustrate their stories. Then help students to make their own picture books, or collect all students' stories into an anthology of student writings entitled *Journey Tales: Migrations of American Wildlife.*

Animal Migration Map

Some animals are year-round residents of their habitat; others are there for only a season or two. Use this map to mark the migration routes of some animals. Use different colors of crayons or markers to draw a line for each animal. Create a simple key by coloring each animal's name with the same color you use on the map.

From *Learning from the Land: Teaching Ecology through Stories and Activities,* Second Edition by Brian "Fox" Ellis. Santa Barbara, CA: Libraries Unlimited. Copyright © 2012.

Animal Migrations

Following is a list of some North American animals that migrate. When you are working with the map, you may choose an animal that is not listed here. Just be sure to choose a specific animal and a specific place on the map, like a Ridley's sea turtle on the Gulf Coast of central Mexico, or a Kentucky warbler that nests along the Scioto River in southern Ohio. The general information given here will stimulate your research, but it won't answer all of your questions.

- Walleye, pickerel, northern pike, and white bass migrate from the Great Lakes into their tributaries (rivers) every spring to spawn.

- Sea turtles migrate to remote beaches along the Florida and Mexico coasts to lay their eggs.

- Wood ducks that spend their summers on the Illinois River spend their winters in the bayous of Louisiana and Arkansas.

- Bald eagles that spend their summers in central Canada spend their winters along the upper Mississippi River. Eagles that spend their summers along the Alaskan coast may spend their winters along the Washington coast.

- Some sandpipers and other wading birds migrate all the way from the northern tundra of Canada to Tierra del Fuego at the tip of South America!

- Gray whales winter in their calving grounds in Baja Mexico and summer off the coast of northern California, Oregon, and Washington, as well as southern Canada.

- Warblers from the forests of Maine, New York, Pennsylvania, Virginia, and North Carolina spend their winters in the tropical rain forests of Costa Rica and Brazil.

- Monarch butterflies fly 1,000 to 2,000 miles from the central and eastern United States to the mountain forests of central Mexico!

- Rocky Mountain elk migrate to mountain ridges in the summer in search of food and down into the valleys in the winter to escape extreme cold and harsh winds.

- The black brandt, a type of goose, migrates 3,000 miles nonstop over the Pacific Ocean, from the Aleutian Islands to the coast of Baja Mexico.

- Rattlesnakes in Texas and other parts of the southwest migrate to a specific cave in their region, where hundreds of snakes hibernate together. In the spring they spread out to fill the surrounding countryside.

- European eels hatch off the coast of Florida. Over the course of three years they use the Atlantic Gulf Stream to make their way to the streams of Europe. After they grow into adults they make the journey back to the Caribbean to spawn.

- Caribou migrate into the boreal forest of northern Canada in the autumn to escape the harsh winds blowing off the open tundra of the Arctic Circle.

- Each fall Ferruginous hawks migrate down the spine of the Appalachian Mountains from Canada and the northeastern United States into Mexico to spend the winter.

Bibliography for Further Research

Berkes, Marrianne, illus. by Jennifer DiRubbio. *Going Home: The Mystery of Animal Migration*. Dawn Publications, 2010. 978-1-58469-126-6. Each two-page spread includes a poem, a brief paragraph of prose about their migration, and an illustration of the creature, often with a map or geographic overlay. All of this is excellent fodder for inspiring student stories.

Great Migrations, The National Geographic Channel seven-part miniseries. This documentary is the best thing I have seen on animal migration, including stunning film footage and the latest science. With careful prescreening, teachers could show parts of the film to introduce different creatures, why animals migrate, and the geography or mapping of migration patterns. Pieces of the film are available on their Web page: http://channel.nationalgeographic.com/channel/great-migrations; where you also find teacher resources, games, photos, excerpts from the book, and lots of ads.

The Seed: A Story of Seasonal Adaptation

And for this end these silken streamers have been perfecting themselves all summer, a perfect adaptation—a prophecy not only of the fall, but of future springs. Who could believe in prophecies that the world would end this summer, while one milkweed with faith matured its seed?

—Henry David Thoreau, from
The Dispersion of Seeds

Comments to the Teacher

THIS IS A STORY OF the seasons.

In part, the story is about the ways in which plants and animals adapt to or are affected by changes in climate. Before telling the story it may be helpful to talk about ways in which people change clothes, eating habits, and thermostats in an effort to adapt to changes in their own climate. This story is also about life cycles, conception, birth, maturation, and death. Notice that there is a wide representation of life forms in the story: a tree, a perennial plant, an insect, an amphibian, a bird, and both large and small mammals. After telling the story, discuss the life stages of these creatures.

To involve the audience, use sound effects and audience participation as you tell the story. For example, invite your listeners to make wind noises whenever you mention the wind, or ask them to make honking noises when you say the word *geese*. Teach your listeners what noises to make and what their cues will be. Then, as you tell the story, use a hand signal to remind your listeners to make the right sound effects at the right time.

And now the story . . .

The Seed

Autumn

A teardrop-shaped seedpod cracks open in the wind. The milkweed scatters its seed. Parachutes catch the breeze. Some seeds fall close by. Others are carried a mile or more away. As they settle on the earth, they are covered in blankets of red, orange, yellow, and brown. Autumn leaves bury the seeds.

Honk, honk, honk. In long V's the Canada geese fly south for the winter. When they fly together, the front of the V formation slices the air to make flying easier for those coming behind. The lead goose also creates a wave of air with the flapping of its wings. This wave of air, or current, helps to lift and lower the wings of the geese behind. So the geese at the  back of the V are pulled along on currents of air by the geese in the front of the line. When the lead goose grows tired, it slips back to rest, and another goose takes the lead. Taking turns all the way, the geese make their way south. By day, they follow landmarks and are guided by the angle of the sun. By night, they are guided by the moon and stars.

In the autumn, monarch butterflies gather to form huge clouds of orange and black. The clouds grow and grow as more butterflies join them, and then the clouds merge into long, trailing rivers of fluttering wings. Together these winged creatures fly south for the winter. When they rest on a tree, the tree is transformed into a living mosaic of orange and black, a dazzling feast for the eyes. These fragile creatures fly more than 1,000 miles to the fir forests of central Mexico.

A huge black bear, full of the fruits of summer, fat from the fall harvest of nuts, begins looking for a place to sleep. In the open prairie there are few caves. Maybe the bear curls up under a cluster of oaks or climbs down a creek bank, where she finds a cave carved by the moving water.

The frog burrows in the mud at the bottom of a pond. Its body slows down. It stops breathing. Its heart almost stops beating. This is true hibernation. If you were to find a frog in the middle of winter, you would think it was dead. If you held it in your hands, the heat from your hands would warm the frog and it would seem to come to life, slowly.

After the work of a season is done, after the acorns have fallen, the leaves of the oak change color. They seem to catch fire as they become brilliant red, bright orange, bronze, and yellow. The winds blow, and the oak tree drops its leaves.

The fox squirrel gathers acorns. It eats its fill and buries the rest of the nuts in the ground. All through the autumn the squirrel eats as many nuts as it can, and puts some aside for winter when food is hard to find. The squirrel also gathers the falling leaves and makes a nest in the crotch of the tree. A thick layer of oak leaves keeps it warm when the cold winds blow.

The north winds howl and snow begins to fall. The seeds of the milkweed are buried in a blanket of white. They lie through the long winter, waiting for spring.

As the snow falls on the milkweed, the geese are having a party under the warm Mexican sun.

Like true kings and queens, the monarch butterflies winter in the tropics of Mexico. They sip nectar from tropical flowers while the north braces for another storm.

As the snow falls, as the pond turns to ice, the frog hibernates safely, snugly, under the mud.

In her cave or snug spot, the bear enters a deep sleep. Bears do not enter a complete hibernation: they stir occasionally in the depths of winter. In the second week of January, bears throughout this land awake to give birth. Cubs are born. The mother licks them clean and goes back to sleep. You may think that big bears would have big babies, but a bear cub could fit in the palm of your hand. The newborn cubs crawl up their mother's belly and drink her milk. The fruits of summer that were converted to fat are once more transformed, this time into mother's milk. Throughout the winter, while their mother is sleeping, the cubs drink and sleep, sleep and drink.

Through the winter the oak tree stands dormant. Its branches are laden with snow and rattled by the winds. But if you look closely at an oak in winter, if you look 10 times more closely at the branches, you will see the tips of buds, and inside the buds you will find tiny leaves formed in the fall, waiting for spring.

The fox squirrel does not hibernate. When the weather is very cold, the squirrel may sleep for five or six days. Could you imagine sleeping for five or six days? When the fox squirrel wakes up, it has one thing on its mind: food!

Do you think squirrels can remember where they buried all of their acorns? Do you think squirrels can even remember where they buried *some* of those acorns? Not at all. So how do they find their nuts? They use their sense of smell. A squirrel can smell through six inches of snow, three inches of leaves, and three inches of dirt—up to one foot of matter—to find an acorn. On the coldest days, the squirrel may add more leaves to its nest before curling up for another long sleep.

Waking, eating, sleeping. Over and over the squirrel wakes, sniffs around, digs up hidden treasures, eats, and goes back to sleep, waiting for the warm breath of spring.

Spring

A warm southeastern breeze begins to blow, and the snow begins to melt. Spring rains start to fall. The milkweed seed swells, soaking in the moisture. A tiny sprout pushes down, deep down, into the Earth.

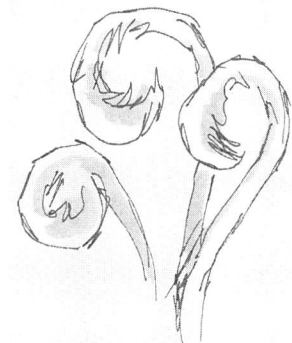

The roots begin to reach out, drinking in moisture and minerals. The leaves push up through the soil, through the clutter of dried stalks and old leaves, reaching for the warmth and light of the sun. Throughout the spring the little plant stretches up toward the sun.

As the days grow longer and the sun returns to this land, so do the geese. Have you heard the geese heralding the return of spring? Honk, honk, honk. Canada geese, which mate for life, build a nest. The female lays eggs, and both mother and father take turns keeping them warm. Both mother and father take turns feeding their young and protecting them from the dangers of this world.

The monarch butterflies also return north. Though one monarch makes the journey south, it takes three or four generations to make the trip north. In early March the butterflies that flew south begin the journey north. When the milkweed of the southern states begin to sprout, these butterflies stop to lay eggs. Their offspring hatch, eat milkweed leaves, metamorphose into adults, and continue the journey. When the milkweed of the central states begin to sprout, these butterflies stop to lay eggs. These grandchildren hatch, eat milkweed leaves, metamorphose into adults, and continue the journey. It is the great-grandchildren of the butterflies that headed south who return to the northern prairie in the late spring, just as the milkweed unfurl their leaves. The monarchs lay their eggs on the young milkweed leaves. The eggs hatch. The young caterpillars eat the leaves. As they eat they grow. As they grow they shed their skin. They eat, they grow, and they shed.

Milkweed leaves are poisonous to most creatures, but the monarch caterpillar is immune. In fact, the poison actually works to help it! After it eats the milkweed, the plant's poison is inside the monarch caterpillar (and later the butterfly). If a creature ate the monarch it would also eat the poison. The

creature would become very sick and vomit or it could die. Any creature that survives would never eat another monarch caterpillar or butterfly!

The bears come out of their den and look for food. The mother has eaten nothing for three months or more. She begins to teach her cubs what to eat, how to hunt and gather. What do you think bears eat? If you said meat, you might be surprised. Black bears usually eat fruits, roots, nuts, and insects. Occasionally they might catch a mouse or chipmunk. Or, if they find a fresh carcass, they might eat some meat. As omnivores they will eat anything that tastes good, but black bears eat mostly plant material.

With the spring thaw the frog begins to stir. It swims to the top of the pond and gulps its first breath as life begins anew! When you hear the frogs singing, you know that spring is truly here.

The frogs are singing to find a mate. The female exudes a string of slimy eggs that the male then fertilizes. This clump of jelly, filled with tiny black dots, hides in the murky water.

If you looked closely at those tiny black dots, 100 times more closely, you would see the stages of evolution before your very eyes. Inside one egg, from a single cell, split in two and split again, comes 4 cells, 8 cells, 16, 32 . . . 64 . . . 128 . . . 1,012 . . . 16,000 . . . 384,000. More than 1 million cells form from those first 2 cells! Cells specialize, change into muscles, skin, gills, nerves. A brain forms.

You went through these same changes inside your mother's womb. You were once an egg. Then you looked like a worm. Then you grew gills, a spine, a brain. You kept growing and changing until you became fully human.

The tadpole forms inside the egg. When it is ready it breaks out of the slimy gelatin and begins to swim around.

As the world bursts into life, the sap rises in the ancient oak. Buds form and flowers burst forth. Oak flowers are like human flowers: the male flower and female flower are separate. That's unusual for a flower. If you cut open a tulip, you will find the pistil (that's the female part) and the stamen (the male part) together in the same blossom. Most flowers are like that. It makes it easier for them to reproduce. But with oak flowers, the male part and the female part are separate. So how do oak flowers get together? They use the wind. The wind carries the pollen from the male flower to the female flower. The pollen fertilizes the ovule (egg), and an acorn gets its start in life.

All that new life is tempting for a starving squirrel. When the weather begins to warm up, the squirrel comes out of its deep sleep. It begins to eat the buds of trees, the eggs of birds, and baby birds. The squirrel cleans out her nest to make room for her own babies, which she must protect from larger enemies. Baby squirrels are so small you could hold several in one hand. They are blind and helpless at birth, completely dependent on their parents. Mother's milk and care help them to grow into frolicking pups.

Summary

A southern breeze blows, and spring spins into summer. The goslings lose their fuzz and begin to grow flight feathers. The adults molt, dropping their old feathers so they can grow new ones. The young geese learn to fly and the old geese prepare for their long flight back to their winter home.

The hot sun showers its light upon the young milkweed. The leaves soak up the sunlight and convert it to food; soon, the plant produces flowers. Tiny buds push up and open, releasing their rich fragrance, attracting butterflies and bees—and humans.

A monarch butterfly visits the pink-and-white blossom to sip the sweet nectar, an enticing treat. The butterfly has pollen on its feet from other flowers it has visited. Look closely at its feet, 1,000 times more closely, and you will see tiny grains of pollen like cockleburs with Velcro grippers. They grab onto the pistil, the female part of the flower, and burrow deep into its base. The pollen joins with the ovary and a tiny seed is made.

Before the butterfly leaves, its feet slip into a little trap on the side of the flower. There it picks up more pollen to take to the next milkweed. The butterfly gets the sweet, sticky nectar as a reward for performing this all-important task of pollination.

This adult butterfly may also lay tiny eggs on the underside of the milkweed's leaves. In about 10 days the eggs will hatch into tiny black, yellow, and white caterpillars. Like generations before, the caterpillar eats, grows, and sheds. It will crack its old skin and crawl out, shedding 12 times before it is ready for metamorphosis. When it is ready, it attaches its back end to a twig and hangs upside down. It does not spin a protective cocoon; instead it splits its skin and rolls it back, exposing a wriggling emerald green sack known as a chrysalis. Using parts of its old self, it dissolves what it used to be to make itself new! After several days, the walls of the chrysalis become transparent. If you look closely you can see the crumpled black and orange wings. The chrysalis cracks open, and the new adult emerges. It pumps fluid into its wings and flies away to repeat the cycle. Three or four generations of butterflies live and die in a prairie summer.

All summer long, the bears forage for food. They eat strawberries in the early summer; blackberries in the middle of summer; blueberries in late summer; and wild grapes, plums, acorns, and hickory nuts in the early autumn. They eat ants from earthen mounds and termites from infested logs whenever they can find them.

Through the warm summer the frog continues to evolve. The tadpole, a fish-like swimmer, sprouts legs. Its tail dissolves and disappears. The frog begins to breathe air and leaps about upon the land. Like a sticky whip, its long tongue snatches insects from the air.

The leaves of the oak tree make food to feed the tree and its acorns. Like food factories, they convert the energy of sunlight into nutrients the tree can use. The little acorns grow plumper and plumper. By late summer, they are fat and brown, full of the nutrients they need to start another small oak—or feed a squirrel.

Scurrying from branch to branch, leaping from tree to tree, playing tag as they race around the trunk, the young squirrels race through summer. Their meals depend on the cycle of seeds, fruits, and nuts. Maple, cherry, mulberry, ash, crabapple, oak, hickory, and walnut—each of these trees will feed the young squirrels in turn. And, the young squirrels help to disperse the seeds.

The leaves of the milkweed continue to harvest the sunlight and convert it to food. The seed pod is formed. The seeds continue to grow, and a parachute begins to form. Each seed is given the food it needs to begin life.

Autumn

Autumn winds blow down from the northwest.

Young frogs swim to the bottom of the pond and burrow under the mud.

The bears begin looking for a place to sleep. During their first winter, the nearly grown cubs will sleep with their mother. Next winter they will have to find a place of their own.

Leaves fall. The days grow shorter, signaling the geese it's time to head south. They fly day and night in long V's, filling the air with their cries. Honk, honk, honk. They herald the return of winter.

The monarch butterflies gather in clouds of orange and black for the long migration south.

The fall harvest fattens young squirrels as they prepare for winter.

A teardrop-shaped seedpod cracks open in the wind. The milkweed scatters its seed. Parachutes catch the breeze. Some seeds fall close by. Others are carried a mile or more away. As they settle on the Earth, they are covered in blankets of red, orange, yellow, and brown. Autumn leaves bury the seeds.

And this story begins again!

Follow-Up Ideas for "The Seed"

Discussion Topic

Ask students how they adapt to the seasons. Invite them to turn to a partner and together discuss one animal in the story. What does that animal do in each season? Ask them to choose a favorite wild animal not in the story, but one that lives in the same ecosystem. What does that animal do when the seasons change?

Tracking the Seasons

❖**Grade Levels**: K–12   **Time estimate**: A lifetime!

❖**Science skills**: Observation; Metric Measurement; Communication; Prediction; Identifying Variables

❖**Objectives**: Help students become more aware of the harbingers of each season!

❖**Materials**: A one-year calendar with large squares to write in; pens or pencils

Instructional Procedures

Introduction: Thomas Jefferson loved the science of phenology, the scientific study of seasonal biological phenomena. Ask students this one simple question: "What season is it and how do you know?"

Activity: Find or create a calendar with large squares so you and the students can record information about each day of the year. Measure and record daily high and low temperatures in Celsius and Fahrenheit. Record on the calendar the day the first robin of spring appears, the date of the last frost of spring (the start of the growing season), the first frost of fall (the end of the growing season), the first date someone in the class has a mosquito bite, the day the maples turn red, the first snowfall, the day the class pet has babies, and so forth. This folk wisdom could be woven into any number of stories.

If the calendar lends itself to use year after year, students in later years can compare the dates on which certain events, like the first snowfall, occur. They could then watch the weather and predict when it will occur this year.

Our third president, Thomas Jefferson, was a huge fan of this kind of science, known as phenology. There are several websites that monitor this information and have gathered hundreds of years of data. This is in large part how we know about climate change. To learn more about these kinds of seasonal changes and to participate in ongoing research visit Project Bud Burst: http://www.budburst.ucar.edu/resources.php.

The National Sustainable Agriculture Information Service has created the mother of all portals for information on phenology with Web links to dozens of sites around the country: http://attra.ncat.org/attra-pub/phenology.html.

Growing a Prairie

❖**Grade Levels**: K–12   **Time estimate**: 2–3 years or class periods

❖**Science skills**: Observation; Prediction

❖**Objectives**: Students will experiment with seed germination in order to grow a small prairie plot.

❖**Materials**: Pie plates for each student; paper towels; stapler and staples; pens or pencils; scissors; seeds of native plants or grasses (available from local greenhouses or through mail order); spray bottle filled with water

Instructional Procedures

Introduction: Many parts of North America were once covered with native prairie plants and grasses—and they can be covered again, when each student grows a minigarden of native plants.

Activity: Have students research perennial flowers and grasses native to your region. Then help them find a place where local genotype seeds can be purchased. To increase germination percentages, students can do additional research on seed preparation: some seeds do better with scarification, scratching the surface; others need to be cold treated, frozen for a few weeks. Obtain enough seed so each student has 10 seeds of 10 different species. Give each student a pie plate or a deep plastic plate as well as 7–10 paper towels.

Students stack the paper towels neatly, then staple the corners to hold them in place. On the top paper towel, students draw or trace an outline of your state.

Using scissors they cut along the outline, cutting through all of the paper towels at once. They should be left with a stack of paper towels cut out in the shape of the state.

Keeping the paper towels in a neat stack, place them in the pie plate. Spray liberally with water until they are soaked. Cover the paper towels with seeds of native plants and grasses. (Sprinkle the seeds onto the towels.) Keep the seeds moist; this may mean spraying them with water several times a day. Keep them in the dark until they germinate. In 7–10 days, the grass will sprout. In 2–3 weeks move them to a windowsill. Your state will once again be covered in native plants!

Students could keep a daily log of observations. They could predict when the seeds will sprout.

With exactly 100 seeds it would be easy to compute germination percentages for each student. The class could also work together to tally germination percentages for each species of plant.

Follow-Up Activities: To find out what you can do to really bring the native plants back to your state, contact a local environmental action committee or visit http://www.prairies.org. One of the best websites to help you plan, plant, and manage a prairie is hosted by Earth Skins Nursery: http://www.earthskinnursery.com/culture.htm.

Sorting Seeds

❖**Grade Levels**: K–12   **Time estimate**: 45–50 minutes

❖**Science skills**: Classification; Reorder, Analyze, and Draw Conclusions

❖**Objectives**: Students will exercise classification skills in determining which seeds are alike and different.

National Standards

Science Standards

NAS 1 Science as Inquiry: Abilities necessary to do scientific inquiry; Understandings about scientific inquiry.

NAS 3 Life Science: Structure and function in living systems; Reproduction and heredity; Regulation and behavior; Populations and ecosystems; Diversity and adaptations of organisms.

Language Arts Standards

NCTE 4 Students adjust their use of spoken, written, and visual language (e.g., conventions, style, vocabulary) to communicate effectively with a variety of audiences and for different purposes.

NCTE 7 Students conduct research on issues and interests by generating ideas and questions, and by posing problems. They gather, evaluate, and synthesize data from a variety of sources (e.g., print and nonprint texts, artifacts, people) to communicate their discoveries in ways that suit their purpose and audience.

❖**Materials**: Various types of seeds and beans (pumpkin seeds, popcorn, sunflower seeds, peas, peanuts, whole cumin, acorns, coriander, whole nutmeg, navy beans, kidney beans, and so forth); bowls or cups; graph paper and pencils; glue

Instructional Procedures

    Introduction: Classification is one of the most basic and important science skills. The following activity works well with seeds, but can also be done with the classroom collection of shoes, a jar of buttons, or any other collection of similar objects.

    Activity: For this simple exercise, collect a wide variety of seeds and beans. You can purchase many kinds of beans and seeds in a grocery store. The easiest way to acquire a variety is to buy a pound or two of 18-bean soup. You'll find an even larger variety of beans and seeds in a health food store. Most hardware stores and nurseries have old seed stock that you may be able to buy at a discount. Feed stores are a good source of bulk bird food. The parks and fields in your area are probably filled with wild grasses, dandelions, milkweed, and other plants whose seeds can be easily collected. If you like, you can gather the seeds yourself. Or, ask your students to help you.

Mix all of the seeds together in a large bowl. Give each student or each co-operative group a cup of mixed seeds. Ask them to sort the seeds in any way that makes sense to them. To encourage creative thinking and problem solving, let the students come up with their own classification systems.

After 10–20 minutes, call the class together to discuss the classification systems they devised and the strengths and weaknesses of those systems. Discuss some of the standard characteristics scientists use in classifying things, such as dispersal strategies (windborne, waterborne, foodborne, burrs); plant types (grasses, shrubs, trees, vines); and dicot (like peanuts) versus monocot (like corn). Then ask students to reclassify their seeds based on the standard scientific method. Have them glue their seed groups onto separate pieces of paper, then produce a chart or graph to show the number of each type of seed in their collection.

Assessment: Their bar graphs can be collected for evaluation of both their math and their logic for grouping seeds.

Follow-Up Activities: These seeds can also be used in a wide range of art projects.

Art Ideas

The leftover Seeds can be used to create a variety of art:

- Bring in a wide variety of seeds and create a mosaic by gluing the seeds to a piece of cardboard. The students can be encouraged to create an image of one animal from the story, a realistic scene, or an impressionistic scene of their favorite season.

- Working in small groups, have each team create a diorama of a different season in an oak savanna prairie or other ecosystem, including seasonal changes and animals and plants in various phases of life.

- Paint a seasonal mural on large sheets of butcher paper.

- Draw the life stages of various animals from egg to adult.

- Create a sculpture of a tree with each of the four seasons represented on a different branch.

- Illustrate the poem or story they write using the Write Your Own Story of the Seasons worksheet (page 000).

Write Your Own Story of the Seasons

Name_____

After listening to "The Seed," fill in the chart below. First, list all of the characters in the column on the left. Then fill in the rows with information about what each character does in each season. Use events described in the story, or use other events that you know about that are not in the story.

   Using the chart, pick one character and write about what that character does in each season. Or, pick one season and write about what all the characters do during that season. (Hint: If you choose a character to write about, look across the chart at what you have written about that character in each of the four seasons. If you choose a season to write about, go down the column to find out what each of the characters does in that season. The short phrases you write in each box could also be used to write a poem.)

CHARACTER	Autumn	Winter	Spring	Summer

Bibliography for Further Research

Johnson, Rebecca. *A Walk in the Prairie*. Carolrhoda Books, 2001. 1-57505-153-2. With a scientific eye and the thoughtfulness of a good teacher, this book is part of a larger series that offers an inspiring immersion into prairie ecology.

Lerner, Carol. *Seasons of the Tallgrass Prairie*. William Morrow, 1980. 13: 978-0688222451. With much more detail, and framed like a prairie textbook for students, this book offers more information than the others, and provides more raw material for creative writing.

McGehee, Claudia. *A Tallgrass Prairie Alphabet*. University of Iowa Press, 2004. 0-87745-897-9. This simple scratchboard illustrated alphabetical list of prairie creatures could be the launch of a whole series of stories about each of the creatures and plants. Each student could choose one and research its habits and habitats to make a new version of the seed.

Bare Stone: A Lesson in Geology and the Successional Stages in the Colonization of a Rock

Let tree roots crack parking lots.

—John Wright, from *Earth Prayers*

Comments to the Teacher

THIS STORY MOVES FROM PERSONAL experience to geologic time. Borrowing from the writing of Aldo Leopold, my goal was to begin "thinking like a mountain," to see each part of the ecosystem as a part of a bigger picture that has existed for millions of years. Hiking up the trail is a literal and metaphoric experience, a leaving of the mundane world and an entering of the allegorical garden. Like a good hike, the story rambles as it covers a lot of ground. The main theme of the story is the successional stages in the gradual colonization of bare stone. The goal is to change listeners' perceptions of the natural world.

Again, it is easy to adapt this story to your personal experience. If you have ever been hiking in the hills or mountains, you have probably seen a tree similar to the one in this story. If it is easier for you to change the story to third person, then change all the personal pronouns to he or she.

Invite the audience to sing a wind chant, "Shoeee, Ah, Oh, Eeee," every time you mention the wind and rain. Every time you say *bare stone*, stomp two times and some audience members will join in. If short on time, the tangents dealing with the Pacific Ring of Fire or Miss Algae can be easily deleted without much loss. Feel encouraged to *tell* this story.

And now the story . . .

Bare Stone

I like to hike. Do any of you like to hike? Where do you like to go? Do you have a favorite hiking trail? [*Allow students to answer your questions, then go on with the story.*]

Think about it: every time you go hiking there is an adventure story in the making!

One of my favorite hikes is in North Carolina, up to the top of Horse Rock Knob in the Black Mountains. The trail is three miles long, and you climb 3,000 feet to get to the top. That's a steep trail! But the view from the top makes it all worthwhile.

One day a few years ago, I decided to take a group of young friends with me. We met at the trailhead early in the morning, before it was too hot, and we started up the trail.

Now, I have three rules about hiking.

Rule number one: No complaining. It brings everybody down. If you have a problem, tell me and we will fix it, but no whining, please. Rule number two: Everyone must help each other. If you are having trouble getting over a fallen log or climbing up a steep rock, do you think the person behind you might be having a hard time? Turn and lend them a hand. It makes the trip easier and more fun. Rule number three: This is the most important one. You must have a good time. Do you want to have a good time? Do you think you can do it? Let's go!

The trail was really steep, especially the first part. We got off to a good start, but pretty soon we were all pretty tired. [*Pantomime this.*] It wasn't long before I heard, "I'm tired," "I'm hungry," "I have to go to the bathroom." I

could understand why they were complaining: that trail was tough!

But I knew something my young friends didn't. I told them, "If you can make it just a little bit farther I know a perfect place to rest, and when we get there we'll have a snack and I'll tell you a story." I'm not sure if it was the story or the snack, but everyone climbed a little bit faster.

We soon came to a perfect resting place along the trail. Everybody who hikes to the top of Horse Rock Knob stops at this spot, where a huge Canadian hemlock tree grows right in the middle of the trail. The roots grow out over a huge rock, and you step on the roots like stairs going up the mountain. The tree leans to one side, and all of its branches point in the same direction, because it has lived its life in the fierce winds of this mountain. You can sit on those roots and rest in the shade of that tree. And as I told you, everybody who hikes this trail stops in the shelter under this tree.

You might be wondering, "What is a Canadian hemlock doing in North Carolina?" Think about this. Several times in the past hundred thousand years,

Bare Stone: A Lesson in Geology

glaciers have pushed down across North America. Glaciers are huge sheets of ice; in some places they were one mile thick. The glaciers moved down from Hudson Bay in Canada. Some came down almost as far as the Ohio River. As the glaciers moved, they pushed all of the cold-weather plants in front of them. When the glaciers retreated, most of the cold-weather plants in warm regions died. But on the tops of mountains, where it's cold, the cold-weather plants survived. That's why, on the tops of the tallest mountains, you'll find balsam fir, Canadian hemlock, and all kinds of flowers and grasses you would not usually find so far south. And that's how a Canadian hemlock came to be growing in the middle of the trail.

All of the hikers were soon sitting underneath that tree. Each of my young friends was sitting on a root, waiting for the story to begin. Close your eyes for a moment and imagine that you are sitting on a root on a rock on the side of a mountain. Can you imagine the textures of root, rock, and soft moss underneath you?

There is a soft breeze blowing, and you hear a Carolina wren off in the distance. [*Whistle.*] Can you smell the fresh scent of balsam in the air? Open your eyes and listen as if you are there.

There was a moment of absolute silence. In that pause, I heard the rocks rumbling behind me, the wind humming in the trees, and right at that moment the tree began to tell me its story.

Take a moment to think about the mountains of North Carolina. Take a moment to remember. Remember a time about one billion years ago. If you think hard enough you might be able to! (The minerals you are made of were there!) Remember a time when the continents of Earth were swimming around. Swimming? That's right!

The native people of North America know that this continent once was an island swimming in the swirl of ocean and sky. Many tribes of North America tell a story about a turtle that carries the world on its back. This is their story: The world was once water; there was no land. The turtle swam down to the bottom of the ocean and brought up mud to make land. Turtle placed this mud on her back to make Turtle Island. You can see the turtle if you look at a map of North America. Florida and Baja Mexico are its hind legs. Central America is its long, curved tail. The land by Hudson Bay and Alaska are its front legs. Okay, so Greenland makes a strangely shaped head, but overall, doesn't it look like a turtle? Native people knew this before there were satellites and space ships and aerial maps. The turtle is a good image for North America. Do you know why? Slowly, like a turtle, North America is swimming. It moves about an inch a year. All of the continents are swimming.

If you take the continents of this world and cut them out and piece them together, they fit together like a puzzle. The fit is not exact, but it's very close. At one time all the continents swam together to form one supercontinent called Pangaea.

Have you ever heard of the Pacific Ring of Fire? Along the western coasts of the Americas is a string of faults and volcanoes. Mount St. Helens and Mount Rainier are volcanoes in Oregon and Washington. Mount Pinatubo is a volcano in the Philippines. All around the perimeter of the Pacific Ocean you will find volcanoes, because the plate of land under the Pacific Ocean is splitting in two. That's right, there's a crack in the Earth! Lava seeps up from the crack near Hawaii and has formed that string of islands.

Think about this: If the continents and the land plates are moving *apart* on one side of Earth, they must be bumping up against each other somewhere else, right? Right! In another part of the world, the subcontinent of India is crashing into Eurasia. As these two land masses collide, the land is pushed up to form mountains. That is how the Himalayas, the tallest mountains in the world, were formed. Mount Everest and K2, both of which are mountains in the Himalayas, are *still* growing! Some say that K2 is growing faster than Mount Everest and it may be the tallest mountain in the world; K2 is growing about 2 inches a year. It's hard to measure a mountain's growth, because it grows so slowly, and even as it grows, wind and rain wear it down.

Let's get back to North America, Turtle Island. Two hundred million years ago, give or take a few hundred thousand years, Europe and Africa crashed into North America and South America. We know this because the rocks in the Americas are the same kind of rocks found in northern Europe and Britain.

When one continent crashes into the other one, one is ground underneath and the other one heaves up. Stones as big as a house were pushed up two miles towards the stars and laid bare. Exposed to the wind and the rain, the stones are worn down. These huge rocks bake in the sun, slowly expanding; then they cool in the long, cold nights, slowly contracting. Crack! Expanding in the heat and contracting in cold, the rocks crack. Over the course of the hot summer, rocks expand; you could measure their growth if you had a huge micrometer. On cold winter nights under the ice and snow, the rocks contract. Crack! Have you ever dropped an ice cube into boiling water? What does it do? It cracks! Have you ever poured hot tea into a cold glass? It shatters in your hand. Well, that's what this rock was doing. Crack, crack! Over many years, cracks opened up and small bits of sand collected in those cracks.

Whoosh! A soft breeze blew, and the small seed or spore of a moss or lichen lodged in the crack. With a little moisture, the spore would sprout.

Do you know the story of lichen? Miss Algae was feeling kind of lonely until she met a Fungi and they took a lichen to each other, but now their relationship is on the rocks! Together algae and fungus create a symbiotic relationship to form a lichen.

Lichens grow incredibly slowly. How slowly do they grow? There was once a scientist who studied lichen when he was a college student. As part of his story, he measured samples of various species of rock lichen to chart their growth. When he was 80 years old, he remeasured those same samples of lichen—and discovered they had only grown a couple of inches! The lichen slowly grew, gripping onto the bare stone. Woo-eesh! The wind and the rain try to wash them off. Woo-eesh! You may want to hold on to your chair, because the wind and the rain may try to wash you off also!

Lichens secrete a little acid, which etches a hole into the rock. This helps the lichen hold on when the wind and rain try to wash it off.

Whoosh! A soft breeze blew a small seed or spore of moss. Moss sprouted in the bit of soil made by the lichen. The moss sent little sprouts in all directions. Roots reached into the cracks, holding on when the wind and the rain tried to blow them off. Woo-eesh! Each spring the carpet of moss grew a little wider. Each autumn it died back and decayed, making soil. Every spring it grew. Every winter it died back. Over hundreds of years, a quarter of an inch of soil formed. On what was once bare stone a community began to grow. There were microbiological organisms so small you would need a microscope to see them. They reproduced by the thousands. Millions of these organisms can live in a cubic foot of good soil. Pill bugs, soil mites, and millipedes moved in.

Have you ever caught a pill bug, the kind that rolls up into a ball when you catch them?

Together these creatures held on—Wooeesh!—when the wind and the rain tried to blow them off.

Whoosh! A soft breeze blew the seed or spore of a fern. In the moisture held by the moss, in the small bit of soil, it sprouted. (Have you seen the fiddlehead fern push up through the earth in the spring, uncurling its long leaf, lifting its fronds to the sun?) Each spring it repeated this dance, growing larger, sturdier, with roots sinking into the soil, into the cracks in the rock. Why? To hold on, because the wind and the rain try to push it off. Each fall the fern died back, adding nutrients to the soil. Over the course of a hundred years maybe another inch of soil formed.

Grasses began to sprout, then wood sorrel and lilies. The community grew. Where you once had bare stone, a village grew, hundreds of insects, dozens of earthworms. A mouse moved in and built a nest under the shade and safety of the fern. The mouse gathered dry moss and grass to build its home. She nursed her newborn offspring. She scouted about for seeds. Maybe she found the cone of a Canadian hemlock and carried it back to her nest. One of the seeds slipped out into the soil. In the moisture and soil made by ferns and mosses, the seed began to sprout, very slowly at first. In 10 years it grew just one foot. Its roots reached into and around the rock to hold on—Wooeesh! —when the wind and the rain tried to wash it off, but it held on. In 30 or 40 years it was as tall as you are. All of its branches grew in the same direction because it lived its whole life in fierce mountain winds.

After 100 or 200 years it was only 20 or 30 feet tall, but its roots had grown thick. They wrap around this rock . . . for you to sit upon . . . to listen to this story . . . Whoosh! And maybe a light breeze blows the seed or spore of a moss or lichen in the crack where root meets rock, and this story begins again.

Discussion Topics

As the story encourages us to do, imagine sitting on a root, growing on a rock, on the side of a mountain with fresh air and bird song. If you were sitting there, what are the stories of the land that we might hear if we tune our ears to listen? Do you have a favorite place in the wild where you like to sit and think?

Thinking about the big ideas, what can we learn from the story of the mountain or the hemlock tree? What job does the mouse or moss or the little creatures do?

A Walk Through Successional Stages

❖**Grade Levels**: K–8   **Time estimate**: 50 minutes

❖**Science skills**: Observation; Metric Measurement; Classification; Prediction; Inference; Identifying Variables; Formulating Hypotheses; Reorder, Analyze, and Draw Conclusions

❖**Objectives**: Students will walk through time, and begin to read the layers of natural history in the landscape.

National Standards

Science Standards

NAS 1 Science as Inquiry: Abilities necessary to do scientific inquiry; Understandings about scientific inquiry.

NAS 2 Physical Science: Properties and changes of properties in matter; Motions and forces; Transfer of energy.

NAS 3 Life Science: Structure and function in living systems; Reproduction and heredity; Regulation and behavior; Populations and ecosystems; Diversity and adaptations of organisms.

NAS 4 Earth and Space Science: Structure of the Earth system; Earth's history; Earth in the solar system.

Language Arts Standards

NCTE 4 Students adjust their use of spoken, written, and visual language (e.g., conventions, style, vocabulary) to communicate effectively with a variety of audiences and for different purposes.

NCTE 7 Students conduct research on issues and interests by generating ideas and questions, and by posing problems. They gather, evaluate, and synthesize data from a variety of sources (e.g., print and nonprint texts, artifacts, people) to communicate their discoveries in ways that suit their purpose and audience.

NCTE 11 Students participate as knowledgeable, reflective, creative, and critical members of a variety of literacy communities.

❖**Materials**: Field guide to wildflowers, graph paper, clipboard, pens or pencils

Instructional Procedures

Introduction: The idea here is for students to begin to read the layers of succession as they walk across an abandoned field. This can be done informally as a simple guided walk or much more formally with field guides, plant identification, and the graphing of their findings then leading to predictions about future succession.

Activity: Find an abandoned field along a forest edge or meadow. Choose a field that has not been mowed in several years. Starting in the middle of the field, slowly walk toward the forest. Pay careful attention to the varieties of plants you see. Challenge students to look carefully for gradual differences in the plants. Walk through the field a second time, pointing out the changes from annuals to perennials, to shrubs and scrub forest, to mature forest plants. Then reverse your direction. As you walk from the forest into the meadow, observe how one plant precedes another, preparing the soil and casting shade for the less sun-tolerant ones. Using a field guide to wild flowers, have students identify the various species of plants they find, then chart their finds on graph paper. Using their findings, have students predict what the field will look like in 5 years, 25 years, and 100 years. Discuss what part of the field and forest has the richest diversity of plants and why. Point out that, given time, the dominant plants of a region will gradually march out of their sanctuaries into any field that is left open to them.

Assessment: This activity is best evaluated based on the level of student participation.

Follow-Up Activities: How can we manage this succession? Challenge students to develop a plan for encouraging the native landscape to regrow into an abandoned field.

Comparing Spores and Seeds

❖**Grade Levels**: K–12   **Time estimate**: 30 minutes

❖**Science skills**: Observation; Metric Measurement; Classification

❖**Objectives**: Students will see the difference between seeds and spores and be introduced to the evolution of plants.

National Standards

Science Standards

NAS 1 Science as Inquiry: Abilities necessary to do scientific inquiry; Understandings about scientific inquiry.

NAS 3 Life Science: Structure and function in living systems; Reproduction and heredity; Regulation and behavior; Populations and ecosystems; Diversity and adaptations of organisms.

Language Arts Standards

NCTE 4 Students adjust their use of spoken, written, and visual language (e.g., conventions, style, vocabulary) to communicate effectively with a variety of audiences and for different purposes.

❖**Materials**: A collection of seeds and spores, graph paper, clipboard, pens or pencils, envelopes

Instructional Procedures

Introduction: In preparation for this study, gather seeds and spores from a natural area. Avoid collecting seeds and spores from endangered plants. Be sure to take less than 10 percent of the available supply of seeds and spores to ensure that there are sufficient numbers for natural reproduction. Place spores from mosses and ferns in one set of envelopes and seeds from grasses and trees in another set of envelopes.

Activity: Give each student or cooperative group a set of envelopes. Ask them to label each envelope with a brief description of what they think it contains. Approach this as you would a guessing game, giving each group time to come to an agreement and to debate the wording of their descriptions.

As a class, discuss what each envelope contains. Explain what a spore is. Also explain the differences between flowering and nonflowering plants. Using a Venn diagram, allow each group to discuss ways in which spores and seeds are alike and different. Continue this discussion as a class. Explain the evolution of flowering and nonflowering plants in geologic time. Using the framework of "Bare Stone," ask students to sort their spores and seeds to show which would come first, second, third, and so on, in colonizing bare stone.

Assessment: Student participation in the discussion is the basis for evaluation. Students could be asked to write a brief definition of seeds and spores using scientific vocabulary.

Follow-Up Activities: Challenge students to look for seeds and spores in their neighborhoods. They could also research online how to sprout each and grow them in their gardens.

Bare Pavement

❖**Grade Levels**: 5–12   **Time estimate**: 60 minutes

❖**Science skills**: Observation; Metric Measurement; Classification; Communication; Prediction; Inference; Identifying Variables; Formulating Hypotheses; Reorder, Analyze, and Draw Conclusions

❖**Objectives**: Students will demonstrate an understanding of how science can influence fiction.

National Standards

Science Standards

NAS 1 Science as Inquiry: Abilities necessary to do scientific inquiry; Understandings about scientific inquiry.

NAS 3 Life Science: Structure and function in living systems; Reproduction and heredity; Regulation and behavior; Populations and ecosystems; Diversity and adaptations of organisms.

NAS 4 Earth and Space Science: Structure of the earth system; Earth's history; Earth in the solar system.

NAS 5 Science and Technology: Abilities of technological design; Understanding about science and technology; Abilities to distinguish between natural objects and objects made by humans.

NAS 6 Science in Personal and Social Perspectives: Personal health; Populations, resources, and environments; Natural hazards; Risks and benefits; Science and technology in society.

Language Arts Standards

NCTE 1 Students read a wide range of print and nonprint texts to build an understanding of texts, of themselves, and of the cultures of the United States and the world; to acquire new information; to respond to the needs and demands of society and the workplace; and for personal fulfillment. Among these texts are fiction and nonfiction, classic and contemporary works.

NCTE 2 Students read a wide range of literature from many periods in many genres to build an understanding of the many dimensions (e.g., philosophical, ethical, aesthetic) of human experience.

NCTE 4 Students adjust their use of spoken, written, and visual language (e.g., conventions, style, vocabulary) to communicate effectively with a variety of audiences and for different purposes.

NCTE 5 Students employ a wide range of strategies as they write and use different writing process elements appropriately to communicate with different audiences for a variety of purposes.

NCTE 7 Students conduct research on issues and interests by generating ideas and questions, and by posing problems. They gather, evaluate, and synthesize data from a variety of sources (e.g., print and nonprint texts, artifacts, people) to communicate their discoveries in ways that suit their purpose and audience.

NCTE 8 Students use a variety of technological and information resources (e.g., libraries, databases, computer networks, video) to gather and synthesize information and to create and communicate knowledge.

NCTE 11 Students participate as knowledgeable, reflective, creative, and critical members of a variety of literacy communities.

NCTE 12 Students use spoken, written, and visual language to accomplish their own purposes (e.g., for learning, enjoyment, persuasion, and the exchange of information).

❖**Materials**: Field guide to plants, graph paper, clipboard, pens or pencils

Instructional Procedures

Introduction: Discuss this simple idea: How long does it take for the native plants and animals to reclaim an urban area? If left to their own devices? If encouraged by people?

Activity: Challenge students to write a story about the successional stages of colonization of an abandoned parking lot. Conduct field research to provide a scientific basis for their stories. If you live in an urban area, explore several lots and compare and contrast the plants that live in each lot. Using a field guide to wild plants, have students identify the various species of plants and chart their findings on graph paper. Based on their observations, have students estimate how long each lot has been abandoned.

Taking this into the realm of science fiction, with an emphasis on science, students could write futuristic stories about the regreening of American cities. Which heroic plants would be the main characters, the colonizers? Which animals would be coconspirators, seed dispersal agents? Which villainous animals would resist the change? What is the conflict and how would they build suspense? Have students edit their stories and tell them to the class.

Assessment: Their stories can be evaluated on both the language arts rubric and how well they use science to support their fiction.

Follow-Up Activities: This could be followed by a discussion of how we can work to make a greener world. Is there an abandoned lot that could be reclaimed as an urban open space?

Bibliography for Further Research

If "Bare Stone" is the story of the mountain ecosystem told by one tree, here are the stories of several other trees:

Cherry, Lynne. *The Great Kapok Tree*. Harcourt Brace Jovanovich, 1990. 13: 978-0152026141.

Reed-Jones, Carol. *The Tree in the Ancient Forest*. Dawn Publications, 1995. 13: 978-1883220310.

Romanova, Natalia. *Once There Was A Tree*. Dial Books, 1985. 13: 978-0140546774.

Vieira, Linda. *The Ever-Living Tree*. Walker Publishing, 1994. 13: 978-0440837428.

"The Ohio" by John James Audubon: In the Wake of Early Naturalists

Neither this little stream, this swamp, this grand sheet of flowing water, nor these mountains will be seen a century hence as I see them now.

—John James Audubon in a letter to
Sir Walter Scott, 1827

Comments to the Teacher

THE JOURNALS OF EARLY NATURALISTS are an inspiring doorway into explorations in local natural history, they serve as models for creative writing, and they can motivate students to conserve and restore the wild places that once existed in their backyard.

Since 1998, I have been touring the country performing various one-man shows as historical naturalists, scientists, poets, and explorers including Charles Darwin, Meriwether Lewis, Gregor Mendel, and John James Audubon. I was surprised at first, but Audubon is by far the most popular because he speaks to artists, historians, and naturalists. He also traveled from Nova Scotia to the Florida Keys, and from Appalachia to the Rockies, keeping a journal everywhere he went. With more than 400 performances under my belt, I have seen the power of stepping into character to help an audience travel back in time and see their world with new eyes and new insights into natural history.

Wherever you live, there are early explorers who kept journals of your local landscape. Research who these people were, learn their stories, share them with your students, and use this great literature as the leaping-off point for preserving or restoring the wild places near your home.

Here is my adaptation of Audubon's essay, "The Ohio," about a flatboat ride he took along the river whose name means *beautiful*.

And now the story . . .

"The Ohio" by John James Audubon

When my wife, my eldest son (then an infant), and myself were returning from Pennsylvania to Kentucky, we found it necessary to purchase a *skiff* to enable us to proceed to our home in Henderson, the river being unusually low. I purchased a large and light boat of that denomination. We procured a mattress, and our friends furnished us with a picnic basket to sustain ourselves. We had two stout servants to row us, and in this trim we left the village of Shippingport, in expectation of reaching our destination in a very few days.

It was in the month of October. The autumnal tints already decorated the shores of that queen of rivers, the Ohio. Every tree was hung with long and flowing festoons of different species of vines, many loaded with clustered fruits of varied brilliancy, their rich bronzed carmine mingling beautifully with the yellow foliage, contrasted with the yet green leaves, reflecting more lively tints from the clear stream than any landscape painter portrayed or poet imagined.

The days were yet warm. The sun had assumed the rich and glowing hue, which at this season produces the "Indian Summer." The moon had rather passed the meridian of her grandeur. We glided down the river, meeting no other ripple of the water than that formed by the propulsion of our boat. Leisurely we moved along, gazing all day on the grandeur and beauty of the wild scenery around us.

Now and then, a large cat-fish rose to the surface of the water in pursuit of a shoal of minnows, which leaped simultaneously from the liquid element. Like so many silver arrows, they produced a shower of light, while the pursuer with open jaws seized the stragglers, and, with a splash of his tail, disappeared from our view. We heard other fishes beneath our bark uttering a rumbling noise, the strange sounds of which we discovered to proceed from the white perch, for on casting our net from the bow we caught several of that species, then the noise ceased for a time.

Nature in her varied arrangements seems to have felt a partiality towards this portion of our country. As the traveler ascends or descends the Ohio, he cannot help but notice that nearly the whole length of the river, on one side it is bounded by lofty hills and a rolling surface, while on the other, extensive plains of the richest alluvial land are seen as far as the eye can command the view. The winding course of the stream frequently brings you to places where the idea of being on a river of great length changes to that of floating on a lake of moderate extent.

(Updated and edited by Brian "Fox" Ellis. To read the original, please visit http://www.foxtalesint.com/Performances/TheOhio.)

Islands of varied size and form rise here and there from the bosom of the water. Some of these islands are of considerable size and value; while others, small and insignificant, seem as if intended for contrast, and serving to enhance the general interest of the scenery. These little islands are frequently overflowed during great floods, and receive at their heads prodigious heaps of drifted timber. We foresaw with great concern the alteration that cultivation would soon produce along those delightful banks.

As night came, sinking in the darkness the broader portions of the river, our minds became affected by strong emotions, and wandered far beyond the present moments. The tinkling of bells told us that the cattle which bore them were gently roving from valley to valley in search of food, or returning to their distant homes. We heard the hooting of a Great Horned Owl, and the muffled noise of its wings as it sailed smoothly over the stream. There was the sound of the boatman's horn, as it came winding more and more softly from afar. When daylight returned, many songsters burst forth with echoing notes, mellow to the listening ear. Here and there the lonely cabin of a squatter struck the eye, giving note of commencing civilization. The crossing of the stream by a deer foretold how soon the hills would be covered with snow.

Many sluggish flatboats we overtook and passed; some laden with produce from the different head-waters of the small rivers that pour their tributary streams into the Ohio; others, of less dimensions, crowded with emigrants from distant parts, in search of a new home. Purer pleasure I never felt; nor have you, kind reader, unless indeed you have felt the like, and in such company.

The margins of the shores and of the river were, at this season, amply supplied with game. A Wild Turkey, a Grouse, or a Blue-winged Teal, could be procured in a few moments; and we fared well, for, whenever we pleased we landed, struck up a fire, and provided as we were with the necessary utensils, procured a good repast.

Several of these happy days passed, and we neared our home, when, one evening, not far from Pigeon Creek (a small stream which runs into the Ohio from the State of Indiana), a loud and strange noise was heard, so like the yells of Indian warfare, that we pulled at our oars, and made for the opposite side as fast and as quietly as possible. The sounds increased, we imagined we heard cries of "murder;" and as we knew that some depredations had lately been committed in the country by dissatisfied parties of aborigines, we felt for a while extremely uncomfortable. Ere long, however, our minds became more calmed, and we plainly discovered that

"The Ohio" by John James Audubon

the singular uproar was produced by an enthusiastic set of Methodists, who had wandered thus far out of the common way for the purpose of holding one of their annual camp-meetings, under the shade of a beech forest. Without meeting with any other interruption, we reached Henderson, distant from Shippingport, by water, about two hundred miles.

When I think of these times, and call back to my mind the grandeur and beauty of those almost uninhabited shores; when I picture to myself the dense and lofty summits of the forests, that everywhere spread along the hills and overhung the margins of the stream, unmolested by the axe of the settler; when I know how dearly purchased the safe navigation of that river has been, by the blood of many worthy Virginians; when I see that no longer any aborigines are to be found there, and that the vast herds of Elk, Deer, and Buffaloes which once pastured on these hills, and in these valleys, making for themselves great roads to several salt-springs, have ceased to exist; when I reflect that all this grand portion of our Union, instead of being in a state of nature, is now more or less covered with villages, farms, and towns, where the din of hammers and machinery is constantly heard; that the woods are fast disappearing under the axe by day, and the fire by night; that hundreds of steamboats are gliding to and fro, over the whole length of the majestic river, forcing commerce to take root and to prosper at every spot; when I see the surplus population of Europe coming to assist in the destruction of the forest, and transplanting civilization into its darkest recesses; when I remember that these extraordinary changes have all taken place in the short period of twenty years, I pause, wonder, and although I know all to be fact, can scarcely believe its reality.

Whether these changes are for the better or for the worse, I shall not pretend to say; but I feel with regret that there are on record no satisfactory accounts of the state of that portion of the country, from the time when our people first settled in it. This has not been because no one in America is able to accomplish such an undertaking. Our Irving and our Coopers have proved themselves fully competent for the task. It has more probably been because the changes have succeeded each other with such rapidity as almost to rival the movements of their pens. However, it isn't too late yet; and I sincerely hope that either or both of them will ere long furnish the generations to come with those delightful descriptions which a country that has been so rapidly forced to change her form and attire under the influence of increasing population. Yes, I hope to read, ere I close my earthly career, accounts from those delightful writers of the progress of civilization in our Western Country. They will speak of the Clarks, the Croghans, the Boones, and many other men of great and daring enterprise. They will analyze the country as it once existed, and will render the picture, as it ought to be. Immortal.

Follow-Up Ideas for "The Ohio"

Discussion Topics

How has the Ohio River changed over time? What can we learn about our local natural history from the journals of early explorers and naturalists? How has the environment changed since the days of the first pioneers?

How can students' journals help them learn science process skills? In what ways can the journals of early naturalists inspire interdisciplinary lessons in art, history, literature, and natural history? These are some of the questions I have pondered as, for the past several years, I have portrayed John James Audubon at schools, nature centers, and museums. These are also questions to be discussed with your students after hearing the story.

Lessons from the Journals of Early Explorers

❖**Grade Levels**: 6–12　　　　　**Time estimate**: 2–3 class periods

❖**Science Skills**: Observation, Inference, Collect and Analyze Data, Draw Conclusions

❖**Objectives**: Students will gain an understanding of the ecological history of their home.

They will see the changes over time as evidenced by the writings of early naturalists.

National Standards

Science Standards

NAS 1 Science as Inquiry: Abilities necessary to do scientific inquiry; Understandings about scientific inquiry.

NAS 3 Life Science: Structure and function in living systems; Reproduction and heredity; Regulation and behavior; Populations and ecosystems; Diversity and adaptations of organisms.

NAS 4 Earth and Space Science: Structure of the earth system; Earth's history; Earth in the solar system.

NAS 5 Science and Technology: Abilities of technological design; Understanding about science and technology; Abilities to distinguish between natural objects and objects made by humans.

NAS 6 Science in Personal and Social Perspectives: Personal health; Populations, resources, and environments; Natural hazards; Risks and benefits; Science and technology in society.

NAS 7 History and Nature of Science: Science as a human endeavor; Nature of science; History of science.

Language Arts Standards

NCTE 1 Students read a wide range of print and nonprint texts to build an understanding of texts, of themselves, and of the cultures of the United States and the world; to acquire new information; to respond to the needs and demands of society and the workplace; and for personal fulfillment. Among these texts are fiction and nonfiction, classic and contemporary works.

NCTE 2 Students read a wide range of literature from many periods in many genres to build an understanding of the many dimensions (e.g., philosophical, ethical, aesthetic) of human experience.

NCTE 4 Students adjust their use of spoken, written, and visual language (e.g., conventions, style, vocabulary) to communicate effectively with a variety of audiences and for different purposes.

NCTE 5 Students employ a wide range of strategies as they write and use different writing process elements appropriately to communicate with different audiences for a variety of purposes.

NCTE 6 Students apply knowledge of language structure, language conventions (e.g., spelling and punctuation), media techniques, figurative language, and genre to create, critique, and discuss print and nonprint texts.

NCTE 7 Students conduct research on issues and interests by generating ideas and questions, and by posing problems. They gather, evaluate, and synthesize data from a variety of sources (e.g., print and nonprint texts, artifacts, people) to communicate their discoveries in ways that suit their purpose and audience.

NCTE 8 Students use a variety of technological and information resources (e.g., libraries, databases, computer networks, video) to gather and synthesize information and to create and communicate knowledge.

NCTE 11 Students participate as knowledgeable, reflective, creative, and critical members of a variety of literacy communities.

❖**Materials**: Paper and pencil, the journals and diaries of early explorers and local pioneers

Instructional Procedures

Introduction: John James Audubon (1785–1851) not only painted all of the known birds of North America—more than 465 species—but he was also an avid writer who published 50 short stories about his travels and travails in the wildest places on the continent. He traveled most of eastern North America, from Newfoundland to New Orleans, and left a treasury of observations. He kept a daily journal with field notes, and later published the five-volume Ornithological Biographies, a compendium of 5- to 10-page bird biographies of all of the birds he painted, along with 50 "Delineations of American Scenery and Manners."

Audubon has inspired me as a traveler and journalist, largely by his great archive of field notes and journal entries. I always wonder as I wander, what did this place look like 100 or 200 years ago? What kind of wildlife might I have seen before the impact of modern life? What would it have been like to camp here a long time ago? Because Audubon was both an artist and a scientist and had an eye for detail, his words and images transport one back to that time, re-creating the ambience of the places he visited. His journals, written in the first half of the 19th century, immerse one in the wilderness as he saw it then.

This passage is from his essay "The Prairie," written in 1811 after a "march of long duration" through the endless grasslands of Illinois, where I now live:

The weather was fine; all around me was as fresh and blooming as if it had just issued from the bosom of Nature. Although well moccasined, I moved slowly along attracted by the brilliancy of the flowers, and the gambols of the fawns along their dams, to all appearance as thoughtless of danger as I felt myself.

My march was of long duration; I saw the sun sinking below the horizon long before I could perceive any appearance of woodland, and nothing in the shape of man had I met with that day. The track which I followed was only an old Indian trace and as darkness overshadowed the prairie I felt some desire to reach a copse in which I might lie down to rest. The Night Hawks were skimming over and around me, attracted by the buzzing wings of beetles which form their food, and the distant howling of wolves gave me some hope that I should soon arrive at the skirts of some woodlands.

Later in the same essay he writes:

Will you believe, good-natured reader, that not many miles from the place where this adventure happened, and where fifteen years ago, no habitation belonging to civilized man was expected, and very few ever seen, large roads are now laid out, cultivation has converted the woods into fertile fields, taverns have been erected, and much of what we Americans call comfort is to be met with? So fast does improvement proceed in our abundant and free country.

Although Illinois is still known as the Prairie State, to walk all day across endless Illinois prairie is unimaginable today. Only 0.01 percent of the virgin prairie remains; and while efforts to preserve what is left and, more importantly, to restore large tracts of prairie, appear to be succeeding, to envisage the prairie landscape as it was, we must rely on the writings of early explorers like Audubon.

The historic perspectives of early explorers and naturalists highlight the ways the landscape has changed over time and can inspire and inform students' writing. Further, their journals can help students become better observers of the world around them. Journaling teaches thinking skills, science process skills, and the mental habits of asking good questions and looking for answers. Ideally, these historic essays, coupled with field ecology and journaling, can both motivate students to take action to improve their local environment and provide a foundation of hard science for such habitat improvement.

What did your local ecosystem look like 100 to 200 years ago? If you could step back in time, what would you see? If you could step into the future, might you see the restoration of some of the natural splendor that once existed? What can you and your students do to begin to re-create a healthy ecosystem for the future?

Activity: Lessons emerge from the past

Elements of these lessons can be easily adapted for almost any grade level from upper elementary through high school. The following steps outline a unit that begins with reading the journals of early explorers and naturalists in your region and leads toward taking action to enhance or restore a local natural area.

There are at least five solid sets of lesson plans that emerge from this exercise:

1. Research the journals of early explorers who may have lived or worked in your region and share these essays and stories with your students.

2. Arrange an outdoor adventure and teach students how to take good field notes.

3. Challenge students to write an essay, poem, or story about the plants and animals they observed, with an emphasis on the interdependence of these organisms.

4. Have students compare and contrast their observations with those of historical authors.

5. Encourage students to draw inspiration from this information to make a difference in their community.

Research the Journals of Early Explorers and Share these With Your Students

For teachers who live in the eastern half of North America, Audubon's journals would be a great place to start researching journal entries that pertain to your area. If Audubon did not visit your area, other historical naturalists probably did and may have written eloquently about what they found:

- The journals of Henry David Thoreau offer a window into rural life and natural history in mid-19th-century New England.

- William Bartram's *Travels* is a classic account of the colonial period in the southeastern United States by one of the world's best botanists.

- John Wesley Powell's journals of his travels in the southern Rockies are often inspiring, especially the passages about his trip down the wild Colorado River.

- Lewis and Clark are arguably two of the finest natural history writers to travel through the west.

- Alexander Mackenzie, in his work for various fur-trading companies, was one of the first Europeans to explore the northern Rockies, Yukon, and Pacific coast.

- The French explorer Père (Jacques) Marquette wrote eloquently about the Great Lakes and the Mississippi watershed.

- Through her journals and books, Catherine Parr Traill documented the flora and fauna of Upper Canada (Ontario) in the mid-1800s.

- La Vérendrye and his sons, from the Trois-Rivières region of Québec, were among the first Europeans to explore the Canadian prairies and Rockies.

- Isabella Bird's "A Lady's Life in the Rocky Mountains" recounts her adventures touring the Colorado Rockies alone on horseback in 1873.

- John Muir's writings about his travels in California and Alaska are undoubtedly some of the best natural history ever written.

- Although not historical authors, Sigurd Olsen, Aldo Leopold, Gretel Ehrlich, Annie Dillard, Barry Lopez, Barbara Kingsolver, Rachel Carson, and John McPhee are writers that inspire high school students to write well.

Read aloud and discuss with your students a few excerpts from your favorite historical author on natural history. If possible, select passages describing landscapes that are familiar to you and your students. Before you read, encourage students to imagine the world through the eyes of this writer. After reading, ask students to describe what they saw as they listened. Invite them to share their impressions of that place today, if it is one that they have visited. They could also share with a partner their memory of some other wilderness that the passage you read stirs in them.

A few open-ended questions can stimulate discussion. For example, Audubon's description of the prairie makes me curious: Do wolves inhabit Illinois today? Have you seen nighthawks hunt for beetles? How would you describe the relationship between nighthawk, beetle, and prairie? What season is the author describing and how do you know? How many miles do you think Audubon walked; how many miles could you walk in a day? Is it possible to walk all day today without seeing a tree? Where in North America might we find an endless prairie like the one Audubon described? Such questions model inquiry-based thinking strategies and help to highlight the ideas that you hope students will gain from these readings.

A more challenging, and therefore more exciting, option is to research the author, learn some of the material by heart, and give a costumed performance as that character. (Please see Stepping Into Character on page 00). This is not as daunting as it may sound. When I perform as Audubon, I do not memorize long passages of his journals. I have read and reread so much of his writing that it emerges almost as if from him whenever I step into the character. With an old hat or some other simple prop and a change of voice, you can make the material come to life in a manner that reading aloud cannot. (As an optional follow-up activity, have students develop costumes and scripts to use in performing the work of an historical author.)

Plan an Adventure and Go Outdoors!

Whether for a hike around the school grounds or a week-long wilderness expedition, take your students outside! Simply sitting for an hour under the oldest tree out behind the school or walking around the block will help them begin to notice the diversity of insects, birds, small mammals, and plants in their environment. Visit the same place at different times and in different weather conditions. Of course, a week-long expedition to a remote wilderness location is guaranteed to inspire great writing, but you might be surprised at the results of a walk through an urban neighborhood. I once led a teachers' workshop in downtown Detroit where the only piece of "wild" we could find was a one-meter-square hole in the pavement. In it were a dead tree, an amazing array of insects, mosses, and lichen, and an abandoned bird's nest—enough to inspire the 20 or so teachers to write poetry and essays about the miracles found in small places. The microcosm of this hole in the pavement became a window into the macrocosm of Detroit's natural history.

Take good notes. Observation is one of the most basic, and therefore most important, skills in science, writing, and art; and taking good notes on what one observes is the essence of journal and nature writing. Have students write down everything they notice. Encourage them to make field sketches—as Audubon did—of what they see and to label the parts. Ask them to describe their sketches. Encourage them to ask questions and look for answers: What is happening? Why? Challenge them to hypothesize about their observations: Why do they think this is doing that? Have students draw a line down the middle of a page and on one side write what they see, the observable facts, and on the other side write questions about those facts and then try to find answers to the questions.

Count things. For example, recently I went kayaking and counted 11 deer—9 does and 2 bucks. I also saw 5 beavers, 2 otters, 2 belted kingfishers, 15 red-tailed hawks, 2 broad-winged hawks, and 1 immature bald eagle. As an isolated tally, my counts may not mean much, but over time such counts give scientists critical information about changes in species populations, ranges, and diversity. Otters did not inhabit the river 10 years ago, but I saw 2 of them. Deer and beaver were completely eliminated in Illinois by the end of the 19th century, but, as my counts show, both have made great comebacks. This habit of observing, counting, and recording can be applied to real scientific research by having students participate in annual bird counts and frog counts, help chart migrations, and measure local diversity through a variety of online environmental monitoring projects.[1]

Always have students note the time of day, weather conditions, temperature, wind direction and speed, phases of the moon, length of day, and any other observable environmental conditions. Thomas Jefferson impressed on Meriwether Lewis the importance of noting weather conditions, temperature, cloud cover, and what the plants and animals were doing in relation to the season. Jefferson himself kept elaborate notes for 30 years on the seasonal changes he observed at Monticello, his estate near Charlottesville, Virginia. The study of such periodic biological phenomena—phenology—helps us make connections between such things as day length and flowering cycles, when birds molt, migrations, hibernation, and other events.

Good notes are the basis of good writing.

Write an Essay, Poem, or Story

Have your students turn their notes into works of prose or poetry in the style and tradition of the American nature writers. From Henry David Thoreau to Annie Dillard, Walt Whitman to Rachel Carson, the approach of these writers has been to use a personal narrative format to draw readers into the wild world. Their narratives blend a poetic appreciation of nature with hard science to stretch the bounds of creative nonfiction. They celebrate the natural world, while raising important issues about our bond to this world.

Using a personal narrative format, students can simply describe what they saw, did, and thought about the outdoor experience. In this way they can create insightful literature that explores their relationship with their local ecosystem. Allow students to write about other personal experiences that they have had outdoors. Watching a robin hunt for worms on the school grounds may have reminded them of a time they saw a territorial dispute between two robins in their backyard. This could lead to an essay about personal space and carrying capacity, about how urban environments challenge our sense of boundaries and create stress for the animals who share our backyards. A simple question that stimulates this kind of deep thinking is, "What can we learn from this animal about being better humans?"

Sharing their stories with others will motivate students to revise and improve their work. It could also stimulate deeper contemplation of their observations. During the writing process, have students share rough drafts with a partner for feedback. Later, encourage them to publish or otherwise share their stories with classmates, other classes, local newspapers, or magazines.

Compare Past and Present

Have students compare and contrast their observations with the perspectives of historical authors. John McPhee's modern essay "Profiles: Travels in Georgia" is one of the best examples of this kind of writing. McPhee, who wrote for the *New Yorker* magazine, retraced the footsteps of William Bartram, a botanist who made some of the first scientific forays beyond the east coast of the United States. McPhee took with him a guide who worked for the Georgia Department of Natural Resources. Making constant references to what Bartram saw 200 years earlier, McPhee leads the reader back and forth through time to challenge us to rethink our relationship to our natural surroundings.[2]

Whether this type of comparison is done on paper or as an oral debate or discussion, the overarching questions are these: How are things different today? How are they the same? What caused the changes? How can we play a role in a healthier future?

Make a Difference

Reading the journals and researching the life stories of great naturalists can be inspiring and instructive. Encourage students to draw inspiration from these explorations to make a difference in their community. This is an important aspect of young people's transition toward adulthood: the eternal internal struggle of hopelessness versus personal empowerment. As teachers, we know that an individual can make a difference and that a class working together can have a global impact.

Gently guiding students as they debate, allow them to generate ideas and a plan of action. Help them find the resources and professional contacts to make their plans a reality. Their plan could be as simple as creating a schoolyard habitat or planting native flowers at home, or as complex as initiating a neighborhood renewal project or raising funds and legislative support for a new nature preserve. (I just this moment got off the phone with my local park district naturalist, negotiating the details of a project to manage prairie restoration along a rails-to-trails project that goes through my neighborhood. We will be working with several schools along the corridor to harvest and germinate prairie seed, to plant and maintain several miles of prairie!)

Late in his life, in a letter to his friend Sir Walter Scott, Audubon observed sadly that he might be the last to see the vast wilderness of North America as it once was. Let's do more than just hope that he was wrong; let's do our part to empower our students to conserve and restore wild places for their children.

Stepping Into Character

Five Easy Steps to Success for Presenting First-Person Interpretations!

1. Which Characters are you Drawn Toward and Why? Who Fits Like a Glove?

The first and most important step is to find a character that fits. Start with a personality that makes it easy for you to step into their shoes. Look for places where your life overlaps with theirs. The more real it is for you, the more real it is for the audience.

2. Start with Primary Source Documents and then Read for Context

Look for autobiographies, letters, and diaries. The goal is to get a glimpse of their world through their words, their ideas, and their use of language, "from the horse's mouth." A children's picture book or grade school biography is a great starting point, because this author has done some of your homework in both boiling down the important points *and* providing a solid bibliography of recommended books at the back. As you read, follow up on threads that intrigue you, and look for exemplary stories, that is, pivotal moments in the character's life that are both a good story and able to give us insight into how the character worked. Read what other scholars have written and read contextual history so you can speak intelligently about this person within his or her historical frame. A rule of thumb: the script should be 70 percent their words, 100 percent their beliefs.

3. Create a Gripping Story Line: Start with a Pivotal Moment in their Career or Create a Timeline of their Life or Engage the Audience in a Public Debate or Re-create a Crisis . . .

There are many ways to approach this, and when you are well-read on the character the outline will suggest itself. Examples include: An old woman looking back on her life; a young man on the verge of some great adventure; a father explaining his life and how he works to a son who is following in his footsteps; a re-created press conference in which you anticipate answers from the press and allow the audience to actually ask questions near the end; or . . . whatever frame you create, it needs to include a logical beginning, middle, and end, dynamic stories, and a chance for the audience to be engaged with the material. Period music is also a plus.

4. Keep Costume and Props Simple, but pay Attention to Detail

Sometimes all you need is a hat, a bonnet, an apron, or a pair of wire-rimmed glasses. Let the story convey authenticity, but make sure your shoes fit the bill! Choose props that are more than stage clutter: they should help tell the story. A Chair, A Crust of Bread, and A Book of Poetry!

5. See Every Show as a Rehearsal

Once you have a solid understanding of the character, a clear outline of where you are going and how to get there, and the right shoes and props, see every time you perform as a chance to explore the character *and* adapt the program to fit the audience. Be conscious of how you can use each performance as a chance to improve elements of the program, working one piece at a time. Take risks to keep it fresh.

From *Learning from the Land: Teaching Ecology through Stories and Activities*, Second Edition by Brian "Fox" Ellis. Santa Barbara, CA: Libraries Unlimited. Copyright © 2012.

Reading the Landscape of America:
Poetry as a Walk through Time

❖**Grade Levels**: 5–12   **Time estimate**: Two to three class periods

❖**Objectives**: Observation, Communication, Collect and Analyze Detail, Make Inferences

National Standards

Science Standards

NAS 1 Science as Inquiry: Abilities necessary to do scientific inquiry; Understandings about scientific inquiry.

NAS 3 Life Science: Structure and function in living systems; Reproduction and heredity; Regulation and behavior; Populations and ecosystems; Diversity and adaptations of organisms.

NAS 4 Earth and Space Science: Structure of the earth system; Earth's history; Earth in the solar system.

NAS 5 Science and Technology: Abilities of technological design; Understanding about science and technology; Abilities to distinguish between natural objects and objects made by humans.

NAS 6 Science in Personal and Social Perspectives: Personal health; Populations, resources, and environments; Natural hazards; Risks and benefits; Science and technology in society.

NAS 7 History and Nature of Science: Science as a human endeavor; Nature of science; History of science.

Language Arts Standards

NCTE 1 Students read a wide range of print and nonprint texts to build an understanding of texts, of themselves, and of the cultures of the United States and the world; to acquire new information; to respond to the needs and demands of society and the workplace; and for personal fulfillment. Among these texts are fiction and nonfiction, classic and contemporary works.

NCTE 2 Students read a wide range of literature from many periods in many genres to build an understanding of the many dimensions (e.g., philosophical, ethical, aesthetic) of human experience.

NCTE 4 Students adjust their use of spoken, written, and visual language (e.g., conventions, style, vocabulary) to communicate effectively with a variety of audiences and for different purposes.

NCTE 5 Students employ a wide range of strategies as they write and use different writing process elements appropriately to communicate with different audiences for a variety of purposes.

NCTE 6 Students apply knowledge of language structure, language conventions (e.g., spelling and punctuation), media techniques, figurative language, and genre to create, critique, and discuss print and nonprint texts.

NCTE 7 Students conduct research on issues and interests by generating ideas and questions, and by posing problems. They gather, evaluate, and synthesize data from a variety of sources (e.g., print and nonprint texts, artifacts, people) to communicate their discoveries in ways that suit their purpose and audience.

NCTE 8 Students use a variety of technological and information resources (e.g., libraries, databases, computer networks, video) to gather and synthesize information and to create and communicate knowledge.

NCTE 11 Students participate as knowledgeable, reflective, creative, and critical members of a variety of literacy communities.

❖**Materials**: Paper and pencil, familiarity with the books *Outside Lies Magic* by John R. Stilgoe (Walker, 1998) and *Reading the Landscape of America* by May Theilgaard Watts (Nature Study Guild Publishers, 1975), access to Google Earth, a printer and graph paper

Instructional Procedures

Introduction: One of the keys to understanding the natural history of the land is to go outside and take a walk with all of your senses open, to truly see what is there and ask how, when, and why. Coming from different backgrounds, John Stilgoe and May Watts have given us invaluable keys to unlocking the layers of natural history by teaching how to read the landscape of urban and wild lands, respectively. Though neither is a read-aloud book for your classroom, both books should be read by every environmental educator.

The idea here is one of the more complex lessons in this book: How can a walk around your neighborhood deepen your sense of place, connect you with the wild world and cycles of nature, and give you both inspiration and information to create a better future for your community?

Activity: Inform students that in a few minutes you are going to take a walk around the school, around the block, around the neighborhood. As they walk they are to take notes on everything and anything that gives any clue as to what happened here before, like a natural history detective that is looking for the age and types of trees, signs of ravines or creeks, soil types, architecture, or landscaping that they can use to figure a date on the building.

For example, in my backyard there is a line of Osage orange trees that tells me my yard was once a cow pasture. There is a steep hill made of gravel that tells me that a glacier piled up rocks right here, forming a glacial moraine. Because I hear coyotes howl whenever there is a siren, I think my creek is a corridor for wildlife. Here is the history of my backyard: (Read the poem "History Is Everywhere I Look").

Take your class for a walk.

At the beginning, point out some of the signs of natural history: maybe there is gravel around the school that was once a shallow sea, limestone, or granite that speaks of glaciers; maybe the oldest trees are only 20–25 years old and that says this was a farm field not so long ago; maybe the local highway was built along an Indian trading route that was once a buffalo path to the salt licks 30 miles south.

Quickly transition from fact finder to questions: How old do you think that house is? What about the addition out back? Was that built as a garage or a carriage house for horses? Do you see that round iron ring for tying off your horse on each of those metal poles? What is the oldest tree you see? Is there a general age for most of the trees? What does this tell us about the neighborhood? Imagine this area without any buildings. How would you describe the lay of the land?

Encourage them to ask questions and look for answers, taking notes on the entire process.

If the weather is nice, find a shady place for them to sit and discuss their notes with a neighbor. Open the conversation so the class can share some of their insights and questions. What can we read from the land about the history of our neighborhood? What did you see that stood out? What clues did you notice?

Using the poem "History Is Everywhere I Look" as a model, ask students to write their own poem that interprets the history and natural history of their neighborhood. They can start by making a list of key concepts or discoveries. Then, using a branching web to organize their thoughts, they can brainstorm ideas following thoughts in several directions. This web can be the outline of their poem. Then on a clean sheet of paper, ask students to write a final draft of their poem.

When you return to your class, you can use Google Earth and a sheet of graph paper to make a landscape map of your discoveries.

Give each student a chance to print a full-page image of their city block using Google Earth. For homework they are to make a similar survey of their neighborhood. Offer points for unusual facts or discoveries. Encourage them to talk to their oldest neighbors to collect information about the types of plants, animals, or soils that make up the ecology of the neighborhood. For bonus points they can go to the library and get old maps of their neighborhood and look for old pictures. Ask the librarian if she can help them find soil maps, historical surveys, or any information that would give them a sense of what their neighborhood was like. For example, there are several oak-named streets, Glenwood is the main thoroughfare, and the high school is called Richwoods, all clues that my neighborhood was once a forest here in the Prairie State.

As students bring in their notes from their block, create a bulletin board that is a map of the school district. As you piece together the big picture, what patterns emerge? What questions come up? What holes can be filled in?

Spend a class period discussing the process, the information discovered, and most importantly what can you do to begin to restore the natural history of the neighborhood. From natural landscaping in their yard to green corridors that connect parks, there is much that can be done to improve the quality of life for ourselves and for the other creatures who share our home!

Assessment: Collect their notes and poems from the first hike for a grade. The information from their homework and their contributions to the larger map can also be evaluated for accuracy. The real assessment comes when a second walk through the neighborhood with their grandkids highlights the trees they planted, the creek they restored, and the green space they reserved!

Follow-Up Activities: Chicago Wilderness (www.chicagowilderness.org), a consortium of more than 100 environmental organizations that encircle the south end of Lake Michigan, has done extensive work along these lines, both ground truth-ing satellite imagery to get a better historical picture and envisioning an eco-region plan for the future of the city and its environs.

History Everywhere I Look

I live in a ranch style brick house
in a semi-suburban development
near a strip mall
with two kids
two dogs
and a swing set in the backyard.

Feeling out of sorts with no sense of place
I wandered out into my backyard
to tend to my herb garden,
to put my hands into the earth.
As I bent to pull a few weeds,
an Osage orange fell from the hedgerow
and rolled up beside my feet.
I paused.
I thought.
Where did that come from?

I looked up at the hedgerow
and saw dozens of these ugly green fruits
and my mind reeled on the question:
Where did that come from?

That hedgerow was here before this development.
When it was a fencerow
cattle grazed in what is now my backyard.
I imagined huge heifers chewing cud
as my house faded in the mists of time
and I was standing in a field
that stretched on for miles.
My mind stretched further back.

I thought of the words:
Osage Orange,
from the Osage River,
from the Osage Indians,
Missouri River country.
In 1804 Meriwether Lewis and William Clark
set out from Illinois,
across the Mississippi,
up the Missouri,
in search of a water route to the Pacific,

---

to negotiate trade relations with the Indians
and explore the unknown Louisiana territory
recently purchased
from a financially strapped Napoleon
to help fund his European wars.
Purchased by Thomas Jefferson,
the author of our Declaration of Independence.

Meriwether Lewis was the first botanist
to describe this plant.
He sent a sample down the Missouri to the Mississippi,
up the Ohio, over the Alleghenies to Monticello.
Jefferson may have grown it in his garden.
Eventually it was proven useful as a hedgerow
and planted by farmers throughout the Midwest,
including the farmer
who sold his land to a developer
who built my house . . .
. . . and I'm standing in my backyard again
holding this freshly fallen fruit.

Notes

1. Some of my favorite online research projects for students are Cornell Lab's Project FeederWatch, www.birds.cornell.edu/PFW; Journey North, a migration-monitoring program, www.learner.org/jnorth; and The National Wildlife Federation has collected a number of citizen science projects on one site: http://www.nwf.org/Wildlife/Wildlife-Conservation/Citizen-Science/Citizen-Science-Programs.aspx.

2. McPhee, John, "Profiles: Travels in Georgia," *New Yorker* 49 (April 28, 1973), pp. 44–46; reprinted in *The John McPhee Reader*, ed. William L. Howarth (New York: Farrar, Strauss, & Giroux, 1976).

Bibliography for Further Research

Bartram, William. *The Travels of William Bartram: Francis Harper's Naturalist's Edition*. University of Georgia Press, 1998. 0-8203-2027-7. Not an easy read by any standard, but worth wading through the archaic language to get a detailed glimpse of the wilderness that was Georgia.

Brehm, Victoria and Sharon L. Dean, eds. *Constance Fenimore Woolson: Selected Stories and Travel Narratives*. University of Tennessee Press, 2004. 1-57233-353-7. This is a woman whose adventures, (more than 100 years ago!) would make most modern mountain men tremble and her powers of observations are truly insightful.

Bruchac, Joseph, illus. by Thomas Locker. *Rachel Carson: Preserving a Sense of Wonder*. Fulcrum, 2004. 1-55591-482-9. This poetic biography boils down the essence of her life in an inspiring and thoughtful manner with gorgeous illustrations. The quotes at the back of the book are by themselves worth acquiring and posting on every science classroom bulletin board.

Giono, Jean. *The Man Who Planted Trees*. Chelsea Green, 1985. 13: 978-1860461170. I was so disappointed when I heard that this book was a work of fiction because I have literally fashioned my life around the model presented by the main character, an old man who plants so many trees in the French Alps that he literally changes the climate! I have planted more than 10,000 trees in the greater Peoria Area. A long story to tell, but I have done it and seen others as well. It is also an award-winning animated short film, well worth tracking down for classroom viewing.

Lasky, Kathryn. *She's Wearing A Dead Bird On Her Head!* Hyperion, 1995. 13: 978-0786811649. This book is not of the same category, but is a good read that explains the origin of the Audubon Society: a group of uppity women who were appalled that birds were becoming extinct for fashionable hats organized a national group dedicated to the preservation of birds.

Muir, John. *The Wild Muir*. The Yosemite Association, 1994. 13: 978-1597140935. Selected and edited by Lee Stetson, who has been playing Muir for more than 20 years, these are very tellable stories that give the reader real insight into the ecosystem of California specifically and our relationships to the earth in a universal manner.

Olson, Sigurd. *Songs of the North*. Penguin Books, 1987. 13: 978-0395901397. Everyone needs a listening spot, a place to be still and quiet, and listen to the songs of the earth. Olson found his spot in the great north woods and shares the wisdom he gained in a manner that inspires others to seek out those magic places.

Index

dragonflies and water quality, 73; for ecological history, 206; for effects on evaporation, 45; for fishing trip, 78; for guessing game poems, 150; for inventive stories, 122–23; for making food web, 65; for migration route map study, 164; for monitoring waste, 23; for nonfiction based on field ecology, 144–45; for observing signs of wildlife, 94–95; for origins of food, 70–71; for pictographic stories, 148; for poetry as walk through time, 214–15; for rewriting a traditional folksong, 90; for science influence on fiction, 195–96; for social action and citizenship, 25–26; for sorting seeds, 181; for student-led drama, 87; for student storytelling, 132–33; for student water weight, 40; for study of predators, 67–68; for successional stages, 192–93; for voice of nature observation, 130; for water weight of foods, 43; for web of life, 61; for wild animal observation, 146–47; for writing migration story, 165–66
Laurasia, 8
La Vérendrye, Pierre, 209
Leaves, 10
Leopold, Aldo, 187, 209
Lewis, Meriwether, 201, 208
Lichen, 190
Life cycles, 10–14, 15–19, 158–62, 174–78
Lopez, Barry, 209

Mackenzie, Alexander, 208
Macroinvertabrate field guide, 77
Maggots, 13
Man-made material, 17
Maps, 37–40
Marquette, Père (Jacques), 209
Mastodons, 9
McPhee, John, 209, 211
Mendel, Gregor, 201
Metamorphosis, 56
Metric measurement, as Science skill, 23, 40, 43, 44, 78, 179, 192, 193, 195
Midges, 55

Migration, animal, 155–62, 169; route map study of, 163–65, 168; writing story about, 165–67
Milkweed, 174–78
Milky Way galaxy, 5–6, 7
Mineral cycle, 3
Minerals: in human body, 5; trees and, 10
Monarch butterflies, 174–78
Mosquitoes, 56, 57
Mosses, 15
Mother Nature had a prairie, 93
Mount Everest, 190
Mount McKinley, 17
Mount Pinatubo, 189
Mount Rainier, 189
Mount St. Helens, 189
Mouse, 102, 103, 140–41
Mud, 13, 16
Muir, John, 209

Native Americans, 13–14, 118; tales, 131–32
Nerves, 142
Neutrons, 5, 6
Neutron stars, 6
Nitrogen, 3, 15–16
Nocturnal, 101, 102
Nonrenewable resources, waste monitoring and, 24–25
Nymphs, 56

Oak tree, 10–11
Observation, as Science skill, 25, 37, 61, 65, 70, 78, 94, 106, 110, 128, 130, 131, 144, 146, 165, 179, 180, 192, 193, 195, 205
Ohio River, 202–4
Oil, 17
Olsen, Sigurd, 209
Organisms, 15–16
Owls, 11–12, 140–41
Oxygen, 15

Pacific Ring of Fire, 189
Pangaea, 8, 15, 189
Petroleum product, 18
Photosynthesis, 10
Pictographs, 139, 148–49

Tall tale, 99–105; explained, 99; rattle-snake science, 100–105; Turtle Island, 135; writing and telling, 111–13

Tend fire, 118, 121

"The Ohio" (Audubon essay), 201–4; follow-up ideas for, 205–18

Thoreau, Henry David, 208, 210

Thunderhead, 34

Traill, Catherine Parr, 209

Travels (Bartram), 208

Turtle Island tales, 135, 189, 190

Variables, identify, as Science skill, 25, 43, 61, 65, 67, 73, 78, 108, 146, 149, 179, 192, 195

Venn diagram, 20

Vision quest, 118, 121

Warm-blooded, 101

Waste monitoring, 23–25

Water cycle, 31–36; follow-up ideas for, 37–48

Water pennies, 55

Weather, 9

Wetlands, 51

Whales, 32–33

Whitman, Walt, 210

Wildlife, backyard, 139, 146–47

Winter, 175

Wolf spider, 141–43

Writing: combining, and telling, 22; individual, exercise, 22–23; migration story, 165–67; story, 22; telling story of wildlife and, 144–46

About the Author

BRIAN "FOX" ELLIS, storyteller, author, and educator, has been touring the world since 1980, collecting and telling stories. He has been a keynote speaker and featured workshop presenter at hundreds of conferences ranging from The International Wetlands Conservation Conference to the National Association of Biology Teachers Conference. His presentations are always custom tailored and contain a mix of pedagogy and practice, humor and inspiration. From the New Mexico Academy of Science to the Michigan Reading Teachers Conference, many of the state educational conferences invite him back on a regular basis! He has also published 15 books, including the award-winning children's picture book, *The Web at Dragonfly Pond* (2006). His DVDs are some of the most award-winning children's storytelling DVDs ever produced. He has written 20 musical theater productions as the artistic director of Prairie Folklore Theatre. Fox is a frequent contributor to a wide range of magazines including trade journals, parenting, academic, and general interest magazines. Fox works extensively as a museum consultant with centers large and small, ranging from the Field Museum and Abraham Lincoln Presidential Library to the Macon County Historical Society and Mary Gray Bird Sanctuary. His artist-in-residency programs have won honors from the North Carolina PTA, where he created a year-long position as the Storyteller in Residence for the Charlotte-Mecklenburg school district.

To learn more about booking a performance or workshop, please visit his website www.foxtalesint.com.

About the Illustrator

VIN LUONG is the son of widely known Chinese painter, Siu Hong. Vin was born in Saigon in 1966. He has studied traditional eastern technique with his father and classic European art at the Western Art School in Vietnam. Blending these two styles he creates art that is at once realistic, surrealistic, Asian, and abstract. He has participated in a wide array of solo and group art shows and has also illustrated several books, including *Content Area Reading, Writing and Storytelling* written by Brian "Fox" Ellis and published by Teacher Ideas Press 2009.